T0212251

Processor Microarchitecture

An Implementation Perspective

Synthesis Lectures on Computer Architecture

Editor

Mark D. Hill, *University of Wisconsin*

Synthesis Lectures on Computer Architecture publishes 50- to 100-page publications on topics pertaining to the science and art of designing, analyzing, selecting and interconnecting hardware components to create computers that meet functional, performance and cost goals. The scope will largely follow the purview of premier computer architecture conferences, such as ISCA, HPCA, MICRO, and ASPLOS.

Processor Microarchitecture: An Implementation Perspective
Antonio González, Fernando Latorre, and Grigorios Magklis
2011

Transactional Memory, 2nd edition
Tim Harris, James Larus, and Ravi Rajwar
2010

Computer Architecture Performance Evaluation Models
Lieven Eeckhout
2010

Introduction to Reconfigurable Supercomputing
Marco Lanzagorta, Stephen Bique, and Robert Rosenberg
2009

On-Chip Networks
Natalie Enright Jerger and Li-Shiuan Peh
2009

The Memory System: You Can't Avoid It, You Can't Ignore It, You Can't Fake It
Bruce Jacob
2009

Fault Tolerant Computer Architecture
Daniel J. Sorin
2009

The Datacenter as a Computer: An Introduction to the Design of Warehouse-Scale Machines
Luiz André Barroso and Urs Hölzle
2009

Computer Architecture Techniques for Power-Efficiency
Stefanos Kaxiras and Margaret Martonosi
2008

Chip Multiprocessor Architecture: Techniques to Improve Throughput and Latency
Kunle Olukotun, Lance Hammond, and James Laudon
2007

Transactional Memory
James R. Larus and Ravi Rajwar
2006

Quantum Computing for Computer Architects
Tzvetan S. Metodi and Frederic T. Chong
2006

Processor Microarchitecture: An Implementation Perspective
Antonio González, Fernando Latorre, and Grigorios Magklis

ISBN: 978-3-031-00601-2 paperback

ISBN: 978-3-031-01729-2 ebook

DOI: 10.1007/978-3-031-01729-2

A Publication in the Springer series
SYNTHESIS LECTURES ON COMPUTER ARCHITECTURE #12

Lecture #12

Series Editor: Mark D. Hill, University of Wisconsin

Series ISSN
ISSN   1935-3235   print
ISSN   1935-3243   electronic

Processor Microarchitecture
An Implementation Perspective

Antonio González
Intel and Universitat Politècnica de Catalunya

Fernando Latorre and Grigorios Magklis
Intel

SYNTHESIS LECTURES ON COMPUTER ARCHITECTURE #12

ABSTRACT

This lecture presents a study of the microarchitecture of contemporary microprocessors. The focus is on implementation aspects, with discussions on their implications in terms of performance, power, and cost of state-of-the-art designs. The lecture starts with an overview of the different types of microprocessors and a review of the microarchitecture of cache memories. Then, it describes the implementation of the fetch unit, where special emphasis is made on the required support for branch prediction. The next section is devoted to instruction decode with special focus on the particular support to decoding x86 instructions. The next chapter presents the allocation stage and pays special attention to the implementation of register renaming. Afterward, the issue stage is studied. Here, the logic to implement out-of-order issue for both memory and non-memory instructions is thoroughly described. The following chapter focuses on the instruction execution and describes the different functional units that can be found in contemporary microprocessors, as well as the implementation of the bypass network, which has an important impact on the performance. Finally, the lecture concludes with the commit stage, where it describes how the architectural state is updated and recovered in case of exceptions or misspeculations.

This lecture is intended for an advanced course on computer architecture, suitable for graduate students or senior undergrads who want to specialize in the area of computer architecture. It is also intended for practitioners in the industry in the area of microprocessor design. The book assumes that the reader is familiar with the main concepts regarding pipelining, out-of-order execution, cache memories, and virtual memory.

KEYWORDS

processor microarchitecture, cache memories, instructions fetching, register renaming, instruction decoding, instruction issuing, instruction execution, misspeculation recovery

Contents

CHAPTER 1

Introduction

Computers are at the heart of most activities nowadays. A processor is the central component of a computer, but nowadays, we can find processors embedded in many other components, such as game consoles, consumer electronic devices and cars, just to mention a few.

This lecture presents a thorough study of the microarchitecture of contemporary microprocessors. The focus is on implementation aspects, with discussions on alternative approaches and their implications in terms of performance, power and cost.

The microarchitecture of processors has undergone a continuous evolution. For instance, Intel has shipped a new microprocessor approximately every year in the recent past. This evolution is fueled mainly by two types of factors: (1) technology scaling and (2) workload evolution.

Technology scaling often is referred to as Moore's law, which basically states that transistor density doubles approximately every 2 years. Every technology generation provides transistors that are smaller, faster and less energy consuming. This allows designers to increase the performance of processors, even without increasing their area and power.

On the other hand, processors adapt their features to better exploit the characteristics of user applications, which evolve over time. For instance, in recent years, we have witnessed an extraordinary increase in the use of multimedia applications, which have resulted in an increasing number of features in the processors to better support them.

1.1 CLASSIFICATION OF MICROARCHITECTURES

Processor microarchitectures can be classified along multiple orthogonal dimensions. Here we will present the most common ones.

1.1.1 Pipelined/Nonpipelined Processors

Pipelined processors split the execution of each instruction into multiple phases and allow different instructions to be processed in different phases simultaneously. Pipelining increases instruction-level parallelism (ILP), and due to its cost-effectiveness, it practically is used by all processors nowadays.

1.1.2   In-Order/Out-of-Order Processors

An in-order processor processes the instructions in the order that they appear in the binary (according to the sequential semantics of the instructions), whereas an out-of-order processor processes the instructions in an order that can be different (and usually is) from the one in the binary. The purpose of executing instructions out of order is to increase the amount of ILP by providing more freedom to the hardware for choosing which instructions to process in each cycle. Obviously, out-of-order processors require more complex hardware than in-order ones.

1.1.3   Scalar/Superscalar Processors

A scalar processor is a processor that cannot execute more than 1 instruction in at least one of its pipeline stages. In other words, a scalar processor cannot achieve a throughput greater than 1 instruction per cycle for any code. A processor that is not scalar is called superscalar. Note that a superscalar processor can execute more than 1 instruction at the same time in all pipeline stages and therefore can achieve a throughput higher than 1 instruction per cycle for some codes.

Very-long-instruction-word (VLIW) processors are a particular case of superscalar processors. These processors can process multiple instructions in all pipeline stages, so they meet the definition of superscalar. What makes a superscalar processor to be VLIW are the following features: (a) it is an in-order processor, (b) the binary code indicates which instructions will be executed in parallel, and (c) many execution latencies are exposed to the programmer and become part of the instruction-set architecture, so the code has to respect some constraints regarding the distance between particular types of instructions to guarantee correct execution. These constraints have the purpose of simplifying the hardware design since they avoid the inclusion of hardware mechanisms to check for the availability of some operands at run time and to decide which instructions are issued in every cycle.

For instance, in a VLIW processor that executes 4 instructions per cycle, the code consists of packets of 4 instructions, each of them having to be of certain types. Besides, if a given operation takes three cycles, it is the responsibility of the code generator to guarantee that the next two packets do not use this result.

In other words, in a non-VLIW, processor the semantics of a code are determined just by the order of the instructions, whereas in a VLIW processor, one cannot totally derive the semantics of a code without knowing some particular features of the hardware (typically the latency of the functional units). By exposing some hardware features as part of the definition of the architecture, a VLIW processor can have a simpler design but, on the other hand, make the code dependent on the implementation, and thus, it may not be compatible from one implementation to another.

1.1.4 Vector Processors

A vector processor is a processor that includes a significant number of instructions in its ISA (instruction set architecture) that are able to operate on vectors. Traditionally, vector processors had instructions that operated on relatively long vectors. More recently, most microprocessors include a rich set of instructions that operate on relatively small vectors (e.g., up to 8 single-precision FP elements in the Intel AVX extensions [17]). These instructions are often referred to as SIMD (single instruction, multiple data) instructions. According to this definition, many processors nowadays are vector processors, although their support for vector instructions varies significantly among them.

1.1.5 Multicore Processors

A processor may consist of one or multiple cores. A core is a unit that can process a sequential piece of code (usually referred to as a thread). Traditional processors used to have a single core, but most processors nowadays have multiple cores. A multicore processor can process multiple threads simultaneously using different hardware resources for each one and includes support to allow these threads to synchronize and communicate under the control of the programmer. This support normally includes some type of interconnect among the cores and some primitives to communicate through this interconnect and often to share data and maintain them coherently.

1.1.6 Multithreaded Processors

A multithreaded processor is a processor that can execute simultaneously more than one thread on some of its cores. Note that both multicore and multithreaded processors can execute multiple threads simultaneously, but the key distinguishing feature is that the threads use mostly different hardware resources in the case of a multicore, whereas they share most of the hardware resources in a multithreaded processor.

Multicore and multithreading are two orthogonal concepts, so they can be used simultaneously. For instance, the Intel Core i7 processor has multiple cores, and each core is two-way multithreaded.

1.2 CLASSIFICATION OF MARKET SEGMENTS

Processors also have different characteristics depending on the market segment for which they are intended. The most common classifications of market segments are the following:

- *Servers*: This segment refers to powerful systems in data centers, which typically are shared by many users and normally have a large number of processors. In this segment, computing power and power dissipation are the most important parameters for the users.

- *Desktop*: This term refers to computers used at home or in offices, typically by no more than one user at the same time. In these systems, computing power and the noise of the cooling solution are normally the most important parameters for the users.

- *Mobile*: This refers to laptop computers, also known as notebooks, whose main feature is mobility, and thus, they operate on batteries most of the time. In these systems, energy consumption is the most important parameter for the users, due to its impact on battery life, but computing power is quite important too.

- *Ultramobile*: In these systems, energy consumption is of paramount importance for the users. Computing power is important but secondary with respect to energy consumption. These systems are usually very small to maximize their portability.

- *Embedded*: This segment refers to the processors that are embedded in many systems that we use nowadays apart from computers. These embedded processors are practically everywhere: in cars, consumer electronics, health-care appliances, etc. Their characteristics vary significantly depending on the particular system where they are embedded. In some cases, their computing power requirements may be important (e.g., in set-top boxes), whereas in many others, the cost is the most important parameter since they have minimal computing requirements, and it is all about minimizing their impact on the total cost of the product. Some embedded processors are in mobile systems, and in this case, energy consumption is also of paramount importance.

1.3   OVERVIEW OF A PROCESSOR

Figure 1.1 shows a high-level block diagram of the main components of a processor that is valid for most processors nowadays, in particular, for out-of-order superscalar processors, which represent the most common organization. It also depicts the main phases that every instruction goes through in order to be executed. Note that these phases do not necessarily correspond to pipeline stages; a particular implementation may split each of them into multiple stages or may group several of them into the same stage.

Instructions are first fetched from the instruction cache. Then they are decoded to understand their semantics. Afterwards, most processors apply some type of renaming to the register operands in order to get rid of false dependences and increase the amount of ILP that can be exploited. Then, instructions are dispatched to various buffers, depending on the type of instruction. Nonmemory instructions are dispatched to the issue queue and the reorder buffer, whereas memory instructions are dispatched to the load/store queue, in addition to the previous two. Instructions remain in the issue queue until they are issued to execution. Operands have to be read before executing an instruction, but this can be done in multiple ways, which we will describe in chapters 5 and 6. Afterward, the result is written back to the register file, and finally, the instruction is committed. An instruction

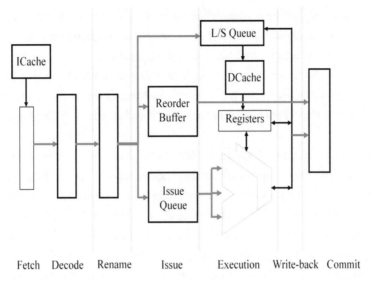

Fetch   Decode   Rename   Issue   Execution   Write-back   Commit

FIGURE 1.1: High-level block diagram of a microprocessor.

remains in the reorder buffer until it commits. The goal of the reorder buffer is to store information about the instruction that is useful for its execution but also for squashing it if necessary.

Memory operations are handled in a special manner. They need to compute the effective address, which typically is done in the same way as an arithmetic instruction. However, besides accessing the data cache, they may need to check their potential dependences with other in-flight memory instructions. The load/store queue stores the requited information for this, and the associated logic is responsible for determining when and in which order memory instructions are executed.

In an in-order processor, instructions flow through these phases in the program order. This means that if an instruction is stalled for some reason (e.g., an unavailable operand), younger instructions may not surpass it, so they may need to be stalled too.

In a superscalar processor, each one of the components described above has the capability of processing multiple instructions at the same time. Besides, it is quite normal to add buffers between some pipeline stages to decouple them and in this manner allow the processor to hide some of the stalls due to different types of events such as cache misses, operands not ready, etc. These buffers are quite common between fetch and decode, decode and rename and dispatch and issue.

1.3.1   Overview of the Pipeline

This section presents an overview of the main components of the pipeline. A detailed description of them is presented in following chapters.

The first part of the pipeline is responsible for fetching instructions. The main components of this part of the pipeline are (a) an instruction cache, where instructions are stored, and (b) a branch predictor that determines the address of the next fetch operation.

The next part is the instruction decode. The main components of this part are decoders, ROMs and ad-hoc circuitry, whose main objective is to identify the main attributes of the instruction such as type (e.g., control flow) and resources that it will require (e.g., register ports, functional units).

Afterward, the instructions flow to the allocation phase, where the two main actions that are performed are register renaming and dispatch. Register renaming entails changing the names of the register operands with the purpose of removing all false dependences and therefore maximizing the instruction-level parallelism that the processor can exploit. This is done normally through a set of tables that contain information about the current mapping of logical names to physical ones and what names are not being used at this point in time, together with some logic to analyze dependences among the multiple instructions being renamed simultaneously, since the destination and source register of a producer-consumer pair has to be renamed in a consistent manner. The instruction dispatch consists of reserving different resources that the instruction will use, including entries in the reorder buffer, issue queue and load/store buffers. If resources are not available, the corresponding instruction is stalled until some other instruction releases the required resources.

The next phase in the pipeline is devoted to instruction issue. Instructions sit in the issue queue until the issue logic determines that their execution can start. For in-order processors, the issue logic is relatively simple. It basically consists of a scoreboard that indicates which operands are available and a simple logic that checks whether the instructions sitting at the head of the queue have all their operands ready. For out-of-order processors, this logic is quite complex since it entails analyzing all the instructions in the queue at every cycle to check for readiness of their operands and availability of the resources that they will require for execution.

After being issued, instructions go to the execution part of the pipeline. Here there is a variety of execution units for different types of operations, which normally include integer, floating-point and logical operations. It is also quite common nowadays to include special units for SIMD operations (single instruction, multiple data, also referred to as vector operations). Another important component of the execution pipeline is the bypass logic. It consists basically of wires that can move results from one unit to the input of other units and the associated logic that determines whether results should use the bypass instead of using the data coming from the register file. The design of the bypass network is critical in most processors since wire delays do not scale at the same pace as gate delays, so they have an important contribution to the cycle time of the processor.

Finally, instructions move to the commit phase. The main purpose of this part of the pipeline is to give the appearance of sequential execution (i.e., same outcome) even though instructions are issued and/or completed in a different order. The logic associated with this part of the pipeline normally consists of checking the oldest instructions in the reorder buffer to see if they have been completed. Once they are completed, instructions are removed from the pipeline, releasing resources and doing some bookkeeping.

Another part of the pipeline that affects multiple components is the recovery logic. Sometimes, the activity done by the processor has to be undone due to some misspeculation (a typical case is branch misprediction). When this happens, instructions have to be flushed, and some storage (e.g., register file) has to be reset to a previous state.

The rest of this document describes in detail the design of the different components of contemporary processors, describing a state-of-the-art design and sometimes outlining some alternatives.

· · · ·

CHAPTER 2

Caches

Caches store recently accessed program data and instructions, in the hope that they will be needed in the near future, based on the observation that program memory accesses exhibit significant spatial and temporal locality. By buffering such frequently accessed data in a small and fast structure, the processor can give the illusion to the application that accesses to main memory are in the order of a few cycles.

Data caches usually are organized in hierarchies of between 1 and 3 levels, so we can talk about the first-level cache, the second-level cache, etc. The lower the level, the closer to the processor the cache is located. Level 1 data caches usually have low associativity and store several tens of kilobytes of data arranged in cache blocks of around 64 bytes each, and they can be accessed in a few cycles (between 1 and 4, usually). Normally, processors have two first-level caches, one for program data (the data cache) and one for program instructions (the instruction cache). Second- and third-level caches are usually between several hundreds of kilobytes and a few megabytes in size, have very high associativity and take tens of cycles to access. These caches usually hold both program data and instructions. Moreover, although each core in a multicore processor has its own (private) first-level caches, higher levels of the memory hierarchy are usually shared among multiple cores. In this section, we will focus the discussion on first-level data caches.

Normally, program data and instruction addresses are *virtual addresses*. Load and store instructions as well as the fetch engine must perform address translation (more on this later) to translate the virtual addresses to *physical addresses*. The caches can be indexed with either virtual or physical addresses. In the former case, the cache access can be initiated earlier, since the address translation can be performed in parallel with the cache access. To deal with potential aliasing problems, the tags are normally generated from the physical address.

Load and store instructions have to take several steps in order to access the data cache. First, the virtual address of the memory access has to be calculated in the address generation unit (AGU), as will be discussed in Chapter 7. In the case of stores, the store queue is updated with this address, and in the case of loads, the store queue is checked for disambiguation (Chapter 6). When the disambiguation logic decides that a load can proceed (or when a store reaches its commit phase), the data cache is accessed.

Instruction cache access is simpler, in that no disambiguation is required since stores to the instruction cache are not allowed. The only requirement is to calculate the virtual address of the instruction to be fetched. This process usually includes *predicting* the next instruction address, a process that will be described in Chapter 3.

The rest of this chapter is organized as follows. First, we explain how virtual addresses are translated to physical addresses. Then we discuss how caches are structured, and we describe several alternatives for cache designs. We focus this discussion on the data cache because it has a more complex pipeline compared to the instruction cache. Finally, we briefly revisit the several alternative cache designs, discussing the implications on the instruction cache.

2.1   ADDRESS TRANSLATION

The *physical address* space is defined as the range of addresses that the processor can generate on its bus. The *virtual address* space is the range of addresses that an application program can use. Virtualizing the program addresses serves two main purposes. First, it allows each program to run unmodified on machines with different amounts of physical memory installed or on multitasking systems where physical memory is shared among many applications. Second, by isolating the virtual address spaces of different programs, we can protect applications from each other on multitasking systems.

The virtualization of the linear address space is handled through the processor's paging mechanism. When using paging, the address space is divided into pages (typically 4–8 KB in size). A page can reside either in the main memory or in the disk (a swapped-out page). The operating system maintains a mapping of virtual pages to physical pages through a structure called the *page table*. The page table usually is stored in the main memory.

It is possible for two virtual pages to map to the same physical page. This is typical, for example, in shared-memory multithreaded applications. When this occurs, the two virtual pages are "aliases" of the same physical entity, so this is called *virtual aliasing*. When implementing the data cache, load/store buffers and all related management mechanisms, it is very important to be able to handle virtual aliasing correctly.

When a program issues a load or store instruction, or an instruction fetch, the page map must be consulted to translate the linear address to a physical address before the memory access can be performed. A linear address is divided into two parts: the page offset and the page number. The page offset corresponds to the N least significant bits of the address—where $N = log_2(PAGE\_SIZE)$—and identifies a byte address inside a page. The page number—the remaining address bits—identifies a page inside the address space.

In a naive implementation, a load instruction (or instruction fetch) would have to perform several memory accesses in order to translate the linear address to a physical one (the page table is in main memory). Since this is a critical operation, all modern processors implement a page table cache, called the translation lookaside buffer (TLB). The TLB is a small hardware cache structure that only caches page table entries.

In some processors, such as the Alpha series of processors, the TLB is entirely software controlled. This means that the operating system has total freedom in how to organize the page map in main memory and also total control of which mappings it wants to be cached to the TLB (there are special instructions to add/remove entries from the TLB).

In the x86 architecture, the TLB is hardware controlled and is mostly transparent to the operating system. In this architecture, the page map has a specific format that the hardware is able to understand. It is the operating system's responsibility to create the page map in a place in the memory where the hardware can find it and in the right format so that the hardware can parse it.

The TLB usually contains in the order of a few tens to a few hundred entries. Associativity may vary, but since it is critical for performance, its access time is usually a single cycle. The TLB is always indexed by the page number of the virtual address, and it returns the corresponding physical page number and some information for this page. The information includes the access rights for the page (if it can be read or written or if it is executable), and if it is mapped to a main memory page or if it is backed up in permanent storage.

2.2   CACHE STRUCTURE ORGANIZATION

A cache consists of two main blocks: the tag array and the data array. The data array stores the application data or instructions, while the tag array is used by the processor in order to match application addresses into data array entries. In Figure 2.1, we can see graphically the cache logical organization.

The data array logically is organized as a group of *sets*. Each set is a collection of *blocks*. The number of blocks in a set is called the *degree of associativity* of the cache. We also say that a cache of associativity N is an N-way associative cache. The i-th cache way is defined to be the collection of the i-th blocks of all sets in a cache. A case with an associativity degree of 1 is called a *direct mapped* cache.

The memory address is split into three parts. The K least significant bits of the address are used to identify which bytes inside a cache block we want to access. This part of the address is called the *block offset*. Assuming the block size is Q bytes, then $K = log_2(Q)$. The next part of the address is called the *index*. As its name denotes, the index is used to identify the position of a set into the data array. For a data cache of S sets, we need $M = log_2(S)$ bits of index.

Different addresses can map to the same set in the data cache (they have the same index), so we need a mechanism to reverse-map indexes to addresses. The tag array serves this purpose. The

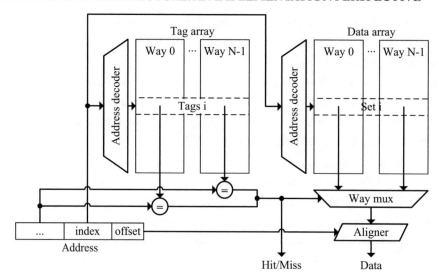

FIGURE 2.1: High-level logical cache organization.

tag array has the same logical organization as the data array (same number of sets and associativity). For each block in the data array, the tag array holds some metadata: the rest of the address bits and the state of the block (valid, etc.)

A memory request accesses both the data and the tag arrays using the index part of the address, but in order to know if the block accessed corresponds to the given address, it must match the rest of the address bits with the tag bits. If the tag bits of the i-th block in the accessed set match, then the correct data is in the i-th block of the corresponding data array set (this is called a *cache hit*). If no tags in the set match the incoming address, then the requested data does not reside in the cache (this is a *cache miss*), and a request to the higher levels of the memory hierarchy must be issued and wait for the data to be installed in the cache before the access can proceed.

2.2.1   Parallel Tag and Data Array Access

The access to the tag and data array can occur in parallel or serially. In the first case (Figure 2.1), a whole set is read from the data array while the tag array is accessed. The address is compared with the tag entries to find in which block of the set reside the data that we are looking for. This information is fed to a multiplexor at the output of the data array (the way multiplexor) that chooses one of the blocks of the set. Finally, the offset part of the address is used to extract the appropriate bytes from the chosen block (this process is called *data alignment*).

The cache access is typically one of the critical paths in a processor; thus, for high-frequency machines, it is pipelined. Figure 2.2 shows a typical pipeline for the parallel tag and data access

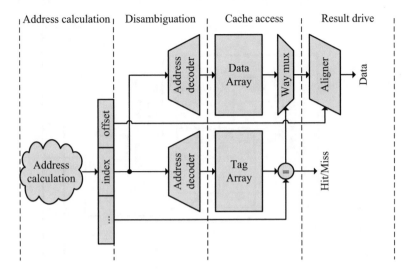

FIGURE 2.2: Parallel tag and data array access pipeline.

corresponding to Figure 2.1. In the pipeline shown, the address decoders of the arrays and the data aligner have been removed from the critical path by placing them in different cycles. There are two critical paths in Figure 2.2. The one is the path that goes through the tag array, the tag comparison and the way multiplexor control. The other is the one that goes through the data array and the way multiplexor data path.

2.2.2   Serial Tag and Data Array Access

An alternative design is one where the tag array is accessed earlier than the data array. The high-level organization of such a design is shown in Figure 2.3. After we access the tag array and perform the tag comparison, we know exactly which of the ways of the data array we must access, so we can change the organization of the cache to take advantage of this fact and remove the way multiplexor. As can be seen in Figure 2.3, the per-way tag comparison signal can be used as a way read/write enable for the data array (shown here as being merged with the decoder output before the array access). This way, the ways of the data array can share the same wires to the aligner. Another benefit of this design is that it has lower energy consumption: the way-enable signal only activates the way where the requested data resides.

Figure 2.4 shows a typical pipeline for the serial tag and data access design corresponding to Figure 2.3. It is easy to see in this figure another very important advantage of this design compared to the one in Figure 2.1. By removing the way multiplexor, we have relaxed the critical paths significantly. First, the data array to way multiplexor data path is entirely removed. Second, the length of

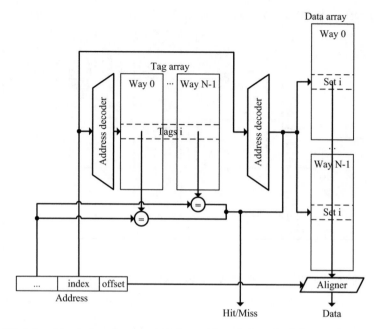

FIGURE 2.3: High-level logical cache organization with serial tag and data array access.

the critical path through the tag array is reduced. This allows this design to run at a higher frequency than the previous one. On the other hand, this design requires one more cycle to access the cache.

It is evident that each design presents different tradeoffs. The parallel tag and data array access design may have lower clock frequency and higher power, but it also requires one less cycle to access the cache. For an out-of-order processor that can hide memory latency, if the data cache is

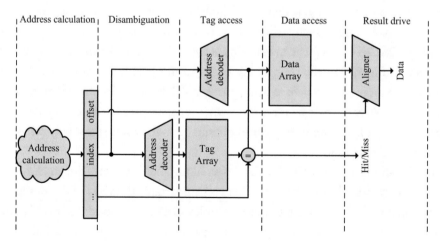

FIGURE 2.4: Serial tag and data array access pipeline.

the determinant factor for the frequency of the processor, it makes sense to implement serial tag and data accesses. On the other hand, for an in-order machine where memory latency is important, it may make sense to access tags and data in parallel.

2.2.3   Associativity Considerations

Direct mapped caches are the fastest to access. In case of parallel tag and data array access, it is the same as Figure 2.2 but without the way multiplexor in the critical path; we do not need the multiplexor because we have only one way. In case of serial tag and data array access, it is the same as Figure 2.4 but without the path from the tag comparison to the data array.

Unfortunately, direct mapped caches suffer more *conflict misses* than associative caches. A conflict miss occurs when N frequently accessed cache blocks map to the same cache set, and the cache associativity is $M < N$. Obviously, the higher the cache associativity, the less conflict misses the cache will suffer. On the other hand, the more ways a cache has, the bigger the way multiplexor becomes, and this may affect the processor's cycle time.

Several proposals exists that attempt to address this situation by modifying the cache indexing function so that an N-way associative cache (where N is small; usually 1 or 2) gives the illusion of a cache of higher associativity [2,37,44].

2.3   LOCKUP-FREE CACHES

When a memory request misses in the first level cache, the request is forwarded to the higher levels of the memory hierarchy. The missing cache access cannot be completed until the forwarded request returns with the data. A *blocking* cache system will stall the processor until the outstanding miss is serviced. Although this solution has low complexity, stalling the processor on a cache miss can severely degrade performance.

An alternative is to continue executing instructions while the miss is serviced. This requires the processor to implement some dependence tracking mechanism (e.g., a scoreboard) that allows the instructions that do not depend on the missing instruction to proceed while the dependent ones are blocked, waiting for the request to be serviced. Such a mechanism of course already exists in out-of-order processors. Another requirement is a *nonblocking* or *lockup-free* cache.

A lockup-free cache is one that allows the processor to issue new load/store instructions even in the presence of pending cache misses. The concept of a lockup-free cache was first introduced by Kroft [28]. In his work, Kroft describes the use of special registers called *miss status/information holding registers* (MSHRs) to hold information about pending misses. Kroft also describes an *input stack* to hold the fetched data until they are written to the data array (this input stack is called the *fill buffer* in modern microprocessors). In the context of lockup-free caches, misses can be classified into three categories:

- *Primary miss*: the first miss to a cache block. This miss will initiate a fetch request to the higher levels of the memory hierarchy [28].
- *Secondary miss*: subsequent miss to a cache block that is already being fetched due to a previous *primary miss* [28].
- *Structural-stall miss*: a secondary miss that the available hardware resources (i.e., MSHRs) cannot handle [10]. Such a miss will cause a stall due to a *structural hazard*.

Several implementations of MSHRs are possible, with different complexity/performance tradeoffs. Here we focus on three such organizations.

2.3.1 Implicitly Addressed MSHRs

This is the simplest MSHR design, proposed by Kroft [28]. This organization is shown in Figure 2.5. As shown in the figure, each MSHR contains the data array block address of the pending misses, along with a valid bit. The block address and the valid bit of an MSHR are set on a primary miss. A comparator also is included, in order to match future misses with this MSHR in order to record all secondary misses of a block in the same MSHR.

Assuming the cache block is divided into N words (typically 32-bit or 64-bit in size), then the MSHR also contains N entries recording miss information. Each entry contains the destination register of the missing instruction and some format information. The format field holds information such as the size (in bytes) of the load instruction, whether the result should be sign or zero extended, the offset inside the word (e.g., for a byte-sized load), etc.

2.3.2 Explicitly Addressed MSHRs

One issue with the implicitly addressed MSHRs is that they can support only one outstanding miss per word. A secondary miss on an already active word field will become a structural-stall miss. The reason for this is that the block offset (the address inside the block) of a missing instruction is implied by the position of the field inside the MSHR.

In figure 2.6, we show another MSHR design proposed by Farkas and Jouppi [10] that improves on the basic MSHR by adding block offset information to each MSHR field. Two misses

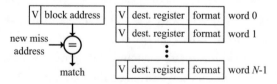

FIGURE 2.5: An implictly addressed MSHR.

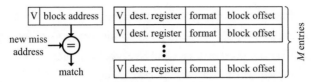

FIGURE 2.6: An explicitly addressed MSHR.

on the same word are allowed to occupy two different entries with the same block offset, inside the same MSHR. This organization also allows decoupling the number of fields from the block size: instead of being forced to have one field per block word, we can have as many as the number of total outstanding misses we want to support per block.

2.3.3 In-Cache MSHRs

An alternative organization that reduces the amount of storage required to hold MSHR information was proposed by Franklin and Sohi [12]. This proposal is based on the observation that the cache block waiting to be filled can serve as the MSHR storage. With in-cache MSHRs, the tag array needs to hold one more bit per cache block, the *transient* bit, to indicate that the block is being fetched. When in transient mode, the tag array holds the address of the block being fetched, and the corresponding data array entry holds MSHR information. The in-cache MSHR can be implicitly addressed or explicitly addressed. The benefit of this design is that we can have as many in-flight primary misses as blocks in the cache.

2.4 MULTIPORTED CACHES

In order to sustain a high execution bandwidth, most modern microprocessors are able to issue up to two load/store instructions per cycle [9,24,26,48]. There are various methods to build a dual-ported cache (a cache that supports two operations per cycle), each with different design tradeoffs. Below we describe a true dual-ported cache design, and then we describe how different commercial microprocessors try to approximate this design.

2.4.1 True Multiported Cache Design

In the true multiported design, all the control and data paths inside the cache are replicated. For a true dual-ported cache, this means that there are two address decoders for the tag and data array, two way multiplexors, double the tag comparators, two aligners, etc. The tag and data array are not replicated, but their internal design is changed to allow two parallel reads to each array cell (allowing two writes per cycle in the cache does not necessarily imply two write ports to the arrays if the processor can guarantee that 1 bit cannot be written twice in a cycle). Although this design provides the highest bandwidth, implementing two read ports to the arrays significantly increases the cache access time, which can affect the processor clock cycle. It is for this reason that no commercial processor that we know of implements a true dual-ported cache.

2.4.2 Array Replication

This design is similar to the true dual-ported design, only, we replicate the arrays as well. Two alternatives are presented: replicating both the tag and data array (i.e., replicating the whole cache) or replicating only the data array and maintaining a true dual-ported tag array. By imitating a dual-ported cache with two copies of a single-ported cache, we can achieve full bandwidth without hurting the processor cycle time. The downside is that this design wastes area, although compared to a true dual-ported design, the area difference is not big. This implementation must keep the two replicas of the data always synchronized, which means that store instructions, cache replacements, block invalidation and similar operations have to be broadcast to both copies of the cache. Several sources [22,35] indicate that the Alpha 21164 microprocessor follows this design, although it is not clear if it replicates only the data array of the entire cache.

2.4.3 Virtual Multiporting

The IBM Power2 [45] and the Alpha 21264 [26] are two examples of processors that implement virtual multiporting to provide dual-ported access to a single-ported cache. Virtual multiporting utilizes time-division multiplexing to perform multiple accesses to a single-ported cache in a single cycle. In particular, the Alpha 21264 double-pumps its data array to provide the illusion that it implements two read ports. In this design, the first load operation accesses the array in the first half of the cycle, while the second load operation proceeds in the second half of the cycle. This technique is abandoned in the latest microprocessor designs because it does not scale. Modern microprocessors that operate at frequencies of several gigahertz cannot afford to double-pump the data array.

2.4.4 Multibanking

Multibanking divides the cache into multiple small arrays (banks), each single ported. Multiple requests can be issued in a cycle if they access to different banks. When two requests map to the same cache bank, we say we have a *bank conflict*. With multibanking, a dual-ported cache still has to have two address decoders for the tag and data arrays, two way multiplexors, two tag comparators, two aligners, etc. Contrary to the true dual-ported cache, though, the tag and data arrays do not have to have multiple ports. When employing multibanking, there are several ways to organize the data in the cache depending on how the indexing function maps addresses to banks, how the cache block

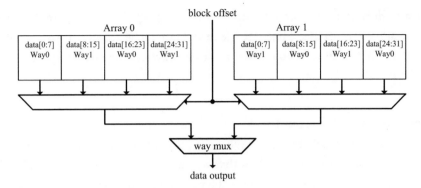

FIGURE 2.7: Arrangement of ways inside a data array bank.

is distributed (or not) among banks, how many banks there are, etc. Multibanking is the preferred method for emulating multiple ports in today's high-performance microprocessors [24].

The MIPS R10000 implements multibanking to allow 2 loads and 1 store access per cycle [48]. The 32-KB data cache of the R10000 is divided into two interleaved banks. The cache is two-way set associative with a block size of 32-B. Each cache bank is logically divided into two logical arrays, and the two ways alternate inside the arrays every 8-B. Figure 2.7 shows the arrangement of block data and ways inside one of the banks. Although we do not have detailed documentation for this, we believe that the least significant bit of the index is used to identify the bank to be accessed.

A different banking organization was chosen by the designers of the AMD Opteron [24]. The Opteron has a 64-KB first-level data cache, two-way set associative with a cache block of 64-B. Each way of the cache consists of eight 4-KB banks. The total cache consists of 16 banks, but a memory instruction accesses one bank of each way for a total of two. Any combination of two load/store requests can issue at the same cycle if they access different banks.

2.5 INSTRUCTION CACHES

Instruction caches have a simpler organization compared to data caches. This is because of the different requirements of instruction fetching vs. program data accesses.

2.5.1 Multiported vs. Single Ported

Program instructions usually are placed consecutively in the main memory, and instruction fetch follows this sequential stream of addresses (except in case of control-flow instructions). To better take advantage of this program property, an instruction cache read usually outputs an entire cache block of data (a data cache read could output as little a single byte if required), which contains multiple consecutive instructions. Thus, a single instruction cache access, on average, feeds multiple instructions to the rest of the machine. Due to this, instruction caches are implemented as single-ported caches.

2.5.2   Lockup Free vs. Blocking

Instruction caches are normally blocking caches, i.e., they are not lockup-free caches. The reason lockup-free caches work for data requests is that we can find future instructions (after the missing load) to execute that do not need the load data. This is not true for instruction fetching. All instructions implicitly depend on their previous in-program order instruction. This is a requirement of correct execution. This means that if we miss in the instruction cache, we must wait for this miss to be satisfied before we can fetch the next instruction. Thus, there is no performance benefit in implementing lockup-free instruction caches.

2.5.3   Other Considerations

An instruction cache with parallel tag and data array access requires one cycle less to fetch instructions compared to one with serial access. This is usually the prefered design because it reduces the penalty of restarting the instruction fetch from the correct address whenever we mispredict the next instruction address. On the other hand, such a design may have lower clock frequency and higher power, so careful studies have to be performed to balance the tradeoffs in each case.

Instruction cache associativity is also another parameter that varies among designs. Similar to data caches, lower associativity caches have lower access time, which allows higher frequency, but higher associativity caches suffer less conflict misses. The sequential access nature of instruction caches produces less conflict misses compared to the random access pattern of data caches, but instruction cache associativity still requires careful tuning of the design.

·   ·   ·   ·

CHAPTER 3

The Instruction Fetch Unit

The instruction fetch unit is the responsible for feeding the processor with instructions to execute, and thus, it is the first block where instructions are processed. The fetch unit mainly consists of an instruction cache and the required logic to compute the fetch address.

High-performance processors can sustain one fetch operation per cycle, which implies that a new fetch address has to be computed every cycle. This means that the next fetch address calculation must occur in parallel to the cache access. However, branch instructions (including conditional branches, jumps, subroutine calls and subroutine returns) introduce a significant extra level of complexity, since the correct fetch address cannot be calculated until we execute the branch.

For this reason, high-performance processors predict the next fetch address. There are two parts in this prediction. The first is predicting the direction of the branch, i.e., taken or not taken. This prediction is performed by what typically is referred to as the branch predictor unit. The second part is predicting the target address of the branch. This prediction is performed by a unit typically called the branch target buffer (BTB). Some processors treat returns from subroutine as special cases and use what is called a return address stack (RAS) unit to predict them.

In Figure 3.1 we can see a high-level block diagram of an instruction fetch unit. There are many alternative fetch unit organizations (with multicycle predictors, multiple prediction levels, etc.), but we believe that the one shown in Figure 3.1 is a straightforward design demonstrating the most important principles of a high-performance and high-frequency fetch unit able to start a new fetch every cycle.

The instruction cache shown in Figure 3.1 accesses the data and tag arrays in parallel, as explained in Chapter 2. We can also see that the instruction TLB is accessed in parallel to the instruction cache. In this design, we assume that the cache arrays are indexed with the virtual address, but the tag matching is done with the physical address. Accessing tags, data and TLB in parallel allows a reduction in the total fetch pipeline depth, which is important for performance (it reduces the cost of restarting the fetch after branch mispredictions).

Figure 3.1 also shows all the relevant next-fetch-address predictors accessed in parallel in the first stage. We show that selecting which predictor output to use for the fetch address takes one more cycle, to emphasize the fact that this pipeline is designed for high frequency. Although this

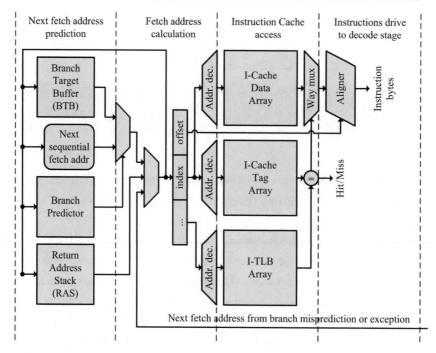

FIGURE 3.1: Example fetch pipeline.

particular design requires four cycles to fetch one block of instructions, since it is fully pipelined, it can start a new fetch every cycle, achieving high throughput.

In the following sections, we describe in more detail the different alternative designs for the instruction cache and the predictors used in modern microprocessors.

3.1 INSTRUCTION CACHE

The instruction cache stores some instructions that are likely to be needed in the near future. It is usually set associative and stores several tens of kilobytes arranged in cache lines of around 64 bytes each. The cache can be indexed with either virtual or physical addresses. In the former case, the access can be initiated earlier since the address translation can be performed in parallel with the cache access. To deal with potential aliasing problems, the tags are normally generated from the physical address.

In superscalar processors, multiple instructions must be fetched per cycle. This is typically achieved by reading consecutive bytes from the cache that are part of the same cache line. In this way, a single memory port may provide the required bandwidth. Once the string of bytes is read, it has to be partitioned into instructions. This is trivial if instructions are fixed size and require some decoding if instructions are variable size.

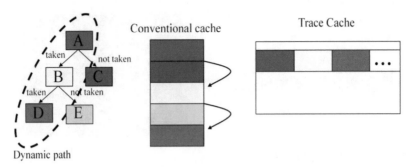

FIGURE 3.2: Conventional instruction cache and trace cache overview.

3.1.1 Trace Cache

Conventional caches store instructions in the same order as they appear in the binary (static order). However, there is an alternative organization that stores the instructions in dynamic order, and it is known as *trace cache* [36]. Figure 3.2 illustrates the key difference between both organizations. There are two key differences between these two organizations: data replication and effective bandwidth per memory port. In a conventional cache, any instruction of a binary appears once at most, whereas in a trace cache, it may appear multiple times, since it can be part of multiple different traces. On the other hand, the maximum bandwidth per memory port of the conventional cache is limited by the frequency of taken branches, which in some integer programs can quite high.

3.2 BRANCH TARGET BUFFER

Predicting whether an instruction is a branch can be done with a hardware table that is indexed with the current fetch address and, for every entry, has as many elements as instructions in a fetch block. A single bit per element may suffice. This bit indicates whether the corresponding instruction is expected to be a branch or not. By adding more bits, one can also predict the type of branch (conditional, call, return, etc.). Once the fetch block is available, the prediction is checked, and in case of misprediction, the table is updated.

To predict the outcome of a branch, the branch unit has to predict its target address, and in case of conditional branches, whether the branch is to be taken.

Most branch instructions encode their target address relative to their location (a.k.a. program counter relative, or PC relative for short) by means of an offset. This is usually the case for branches that correspond to conditional and loop structures in programs, which are responsible for the majority of branches. These branches have two important characteristics: first, their target address is always the same, and it is not very far from the branch instruction. Because of that, the PC-relative encoding is quite appropriate. For these branches, the target address can be computed in the cycle after they are fetched, instead of being predicted. This would just require a dedicated adder in the

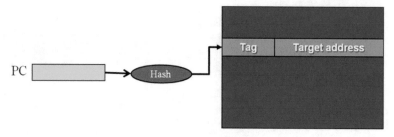

FIGURE 3.3: The branch target buffer.

front-end, but the main drawback is that it would introduce a 1-cycle bubble for every taken branch, which for high-performance processors can be an important penalty.

To avoid this bubble, the target address has to be predicted. This is normally accomplished by having a hardware table that is indexed with the fetch address and has an entry for every instruction in a fetch block. This entry contains the predicted target address if it corresponds to a branch. The prediction is just the target address of the previous time the branch was executed. This is guaranteed to be correct for PC-relative branches and happens to be quite accurate also for computed branches (i.e., those branches whose target address is not known at compile time and is computed at run time). Note that this table can be combined with the one described above for predicting if an instruction is a branch, and the combined table is normally called branch target buffer (BTB). Figure 3.3 illustrates its main structure.

3.3   RETURN ADDRESS STACK

There is a particular type of computed branches that deserve special attention. These are the return-from-subroutine instructions. These instructions have variable target addresses, which depend on where the corresponding call to a subroutine is placed. A BTB could predict them with relatively good accuracy, since often a given subroutine is called multiple times in a row from the same location (e.g., when the call is in a loop). However, there is a special mechanism that is very simple and is even more accurate. It is called the return address atack (RAS).

The RAS is a hardware LIFO structure, where every time the processor fetches a subroutine call, the address of the next instruction is pushed in. When a return instruction is fetched (or is predicted to be fetched by the BTB), the most recent entry of the RAS is popped out and used as the target address prediction for the return instruction. If the RAS had an unbounded number of entries, it would be able to correctly predict practically all return instructions (all if the return address is not explicitly changed inside the subroutine, which is totally discouraged by good programming practices, and it is very rare in typical workloads). In practice, the RAS has a relatively small number of entries (e.g., a few tens). When a call instruction finds it full, the oldest entry is lost, and this will cause a misprediction in a future store. However, this situation is very rare in many programs, since it only happens when the subroutine nesting level is higher than the number of RAS entries, and it has been experimentally observed that the nesting level is rarely higher than a few tens. Obviously, this is not the case for recursive subroutines.

3.4   CONDITIONAL BRANCH PREDICTION

Regarding whether the branch is to be taken, the prediction is a must practically in any processor for conditional branches, since the condition depends on run-time data and cannot be computed until the execution stage of the pipeline. From the time a branch is fetched until it is computed, it may easily take more than 10 cycles in many microprocessors, so waiting for its computation to fetch the next block is not an option in practically any processor.

Branch condition prediction can be done statically (i.e., by the compiler/programmer), dynamically or a combination of both.

3.4.1   Static Prediction

Static prediction can be done with profiling information, by collecting the most frequent outcome of each branch for a particular run and using it as the prediction. Without profiling information, it can be relatively accurate to predict that loop closing branches will be taken. For the rest (e.g., branches corresponding to conditional structures), it is, in general, tough to know a priori which direction they will take.

Static prediction of the condition is very simple from the hardware standpoint; it just requires some bits (one may be enough) in the instruction so that the compiler can encode the prediction. A minimalist form of static prediction is to predict the same outcome for all branches (i.e., all taken or not taken), which avoids the need for any extra bit in the instructions. On the other hand, dynamic prediction requires more complex hardware, but it is, in general, much more effective, so it is present in practically all processors. The advantage of dynamic predictors comes from the fact that they use the actual data of the running application and that they can change the prediction for every dynamic instance of the same static branch.

3.4.2   Dynamic Prediction

Dynamic prediction is based on some hardware that stores past information of the running application and uses this information to predict every branch. A simple and quite commonly used predictor in the past consists of a table that contains 2^n entries of 2 bits each (see Figure 3.4) [40]. The table is indexed with the address (the PC) of the branch instruction, for instance, using the n least significant bits (if instructions are fixed size, the few least significant bits that represent the byte offset are

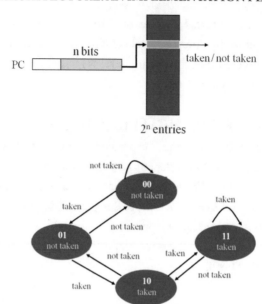

FIGURE 3.4: A local branch predictor.

normally not considered). The corresponding entry is then used to predict the branch condition and is updated with the branch condition outcome once it is available, to reflect the most recent history of this branch. The two-bit entry implements a finite-state machine that is used for making the prediction and storing the recent history. The finite-state machine depicted in Figure 3.4 is often referred to as a two-bit saturating counter.

This predictor is called a local branch predictor since the prediction of each branch is made using the history of the same branch, if we ignore the effect of aliasing. This predictor is designed to work well with highly biased branches. Branches that are almost always taken will tend to be in the "11" state, whereas branches that are almost always not taken will tend to be in the "00" state. This predictor is also good if a branch changes its bias from time to time, as far as it keeps a particular bias for some time.

Since the table has a finite number of entries, sometimes two different branches happen to use the same entry. This is called aliasing and, in general, degrades the accuracy of the predictor. In particular, for this predictor, if the two aliased branches have the same bias, then this aliasing has minimal effect, but if they have opposite bias, the accuracy of their predictions is dramatically compromised.

A 2-bit local branch predictor typically has an accuracy higher than 80% and, for some programs, it can be as high as 99%. This may be adequate for some microprocessors, but for current high-performance processors, a 10% misprediction represents an important penalty. This is mainly due to the high penalty of each misprediction. As described in more detail in Chapter 8, a branch misprediction causes a flush of the pipeline, and the fetch unit is redirected to the correct path. This implies that instructions from the correct path will not be able to reach the execution stage until they traverse the entire pipeline from fetch to execute. In current high-performance processors, this is typically more than ten cycles, so for programs where branches are quite frequent (e.g., 1 every 10 instructions), introducing a bubble of more than ten cycles every misprediction represents a signifi- cant penalty. The penalty of a bubble in the front end depends on the particular microarchitecture (bandwidth of the frontend, bandwidth of the back-end, etc.) and the characteristics of the code, so it cannot be computed without a cycle-level simulator (or the actual hardware). However, note that current superscalar processors have a front-end bandwidth that can be around 4 instructions per cycle, so a 10-cycle bubble in the front end represents a lost opportunity to fetch 40 instructions.

To further reduce branch misprediction penalty, current microprocessors usually include a correlating predictor, also known as two-level branch predictor [49]. A correlating predictor makes a prediction of a given branch using history not only of the branch itself but from other "neighbor" branches too.

A simple and effective way to build a correlating predictor is shown in Figure 3.5 [31]. There is a register, which is called branch global history, that stores the outcome of the most recent branches (1 bit to indicate taken or not taken). Something like 10 to 20 bits of history may be ad- equate. This history is combined with the PC of the branch through a hashing function to generate an index to a table that contains 2-bit saturating counters. The entry is used to make the prediction and is updated with the outcome of the branch in the same manner as for the local predictor de- scribed above. This predictor is called gshare.

The basic idea behind gshare and, in general, all correlating predictors is to try to use a dif- ferent finite-state machine for every different combination of static branch and history. In this way,

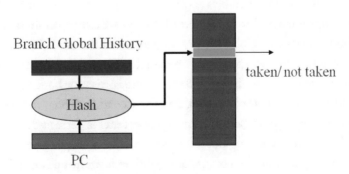

FIGURE 3.5: A gshare predictor.

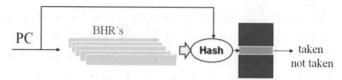

FIGURE 3.6: A correlating predictor with multiple global branch history registers.

the prediction is based on both what this particular branch did in the past and what its "neighbor" branches did. Of course, from the cost standpoint, it is more cost effective not to provide an entry for every potential combination of PC and history and tolerate some degree of aliasing. It has been experimentally proven that a bit-wise exclusive OR between the branch global history and the least significant bits of the PC is a simple and effective hash function in terms of minimizing aliasing.

Correlating predictors vary in the way that global history is used and how it is combined with the identity (i.e., PC) of the predicted branch. For instance, we may have multiple global branch history registers, as illustrated in Figure 3.6. A particular branch uses a concrete history register based on its PC, and this concrete register is hashed with the PC to get the index to the particular finite-state machine that will be used for the prediction.

The accuracy of correlating predictors depends on the amount of global history used, the number of finite-state machines (i.e., two-bit saturating counters) and how the global history and the branch PC are used to determine the concrete finite-state machine to be used in each case. There is no particular configuration that is the best for all codes. For instance, sometimes it is best to have more global history, whereas in other cases, global history does not help and may even cause degradation in accuracy. Because of that, some processors use hybrid branch predictors.

A hybrid branch predictor [31] consists of multiple predictors such as those described above and a selector (see Figure 3.7). The selector is responsible for deciding which of the individual predictors is more reliable for every branch prediction. When the hybrid predictor consists of two individual predictors, the selector resembles another branch predictor since it has to predict a Boolean value. It can be implemented through a table of two-bit saturating counters that is indexed by a combination of the PC and some global history (or just one of the two). The uppermost bit of the counter indicates the preferred predictor. Once the actual outcome of the branch is known, the entry is updated in the following manner. If both predictors are correct, or both are wrong, the counter is not modified. If the first predictor is correct and the second is wrong, the counter is increased, whereas if the first is wrong and the second correct, the counter is decreased.

Hybrid predictors are interesting not only because different types of codes may be more adequate for different predictors but also because of the different warm-up time of the predictors. We call warm-up time of a predictor the time required from an initial state where its different tables and

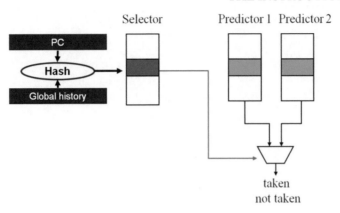

FIGURE 3.7: Hybrid branch predictors.

registers contain irrelevant information (e.g., from another application) to the time when it reaches a steady state regarding prediction accuracy. The warm-up time of a local predictor is quite short; as soon as a bias branch is executed a couple of times, it is predicted correctly. On the other hand, a correlating predictor has a much longer warm-up since a given static branch makes use of multiple finite-state machines, one for each different global history. In order to reach its steady state, all these finite-state machines have to be biased towards its appropriate value, which requires that this branch be executed a number of times proportional to the different values of the global history. Taking into account that every time there is a context switch, the state of the predictor is irrelevant for the new process, the warm-up time may have an important impact in multiprogrammed systems. Because of that, one may consider using a hybrid predictor consisting of a local predictor and a correlating predictor. Right after each context switch, the local predictor would usually be better, due to its faster warm-up, but after some time, the correlating predictor will start to predict better than the local one.

CHAPTER 4

Decode

The purpose of the intruction decode stage is to understand the semantics of an instruction and to define how this instruction should be executed by the processor. It is in this stage that the processor identifies:

- What type of instruction this is: control, memory, arithmetic, etc.
- What operation the instruction should perform, for example, whether it is an arithmetic operation, what ALU operation should be performed, whether it is a conditional branch, what condition should be evaluated, etc.
- What resources this instruction requires, for example, for an arithmetic instruction, which registers will be read and which registers will be written.

Typically, the input to the decode stage is a raw stream of bytes that contains the instructions to be decoded. The decode unit then must first split the byte stream into valid instructions by identifying instruction boundaries and then generate a series of control signals for the pipeline for each valid instruction. The complexity of the decode unit depends heavily on the ISA and the number of instructions that we want to decode in parallel.

In the first section of this chapter, we will briefly explain how decoding works in a RISC machine. In the following section, we will revisit the encoding of the x86 instructions, and we will comment on how each ISA feature affects the decoding complexity. Next, we will discuss the dynamic translation technique that modern x86 processors utilize, to translate x86 instructions to simple RISC-like operations. Finally, we will describe the decoding pipeline of a modern, out-of-order x86 processor that dynamically translates x86 instructions into an internal RISC-like set of instructions.

4.1 RISC DECODING

A typical RISC decoding pipeline can be seen in Figure 4.1. In the figure, we show a superscalar RISC machine that can decode 4 instructions in parallel. For the discussion throughout the rest of this section, we will assume the architecture of Figure 4.1.

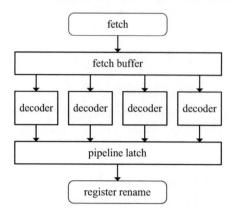

FIGURE 4.1: Typical RISC decode pipeline.

Typically, RISC instructions are simple to decode. Most RISC processors have a fixed instruction length, which makes finding the boundaries of the instructions in the fetch buffer and passing the raw bits of the instructions to the decoders trivial. The only thing that we need is the index inside the fetch buffer of the first instructions (it may not be aligned to the beginning of the buffer), which is easy to obtain from the low-order bits of the PC.

Moreover, RISC ISAs have very few encoding formats, which means that there are very few variations in the position of the opcode and the operands in the instructions. This, combined with the fact that RISC instructions are "simple," which means that they gerenate few control signals for the pipeline, makes the decoders relatively simple.

The simplicity of RISC instructions enables high-performance processor implementations to have single-cycle decoding, using simple PLA circuits and/or small look-up tables. This is, of course, one of the original goals of RISC: to allow easy decoding and simple execution control, to facilitate high-performance implementations.

4.2   THE X86 ISA

The x86 is a variable-length, CISC instruction set. The format of an x86 instruction is shown in Figure 4.2, borrowed from the Intel Architecture Manual [19].

An x86 instruction constists of up to four prefix bytes (optional), a mandatory opcode that can be from 1 to 3 bytes, and an optional addressing specifier consisting of the ModR/M byte and maybe the SIB (scale-index-base) byte. Some instructions may also require a displacement (up to 4 bytes) or an immediate field (also up to 4 bytes).

The instruction prefixes serve several purposes. For example, a prefix can modify the instruction operands: the segment override prefix changes the segment register of addresses, while the operand-size override prefix changes between 16- and 32-bit register identifiers.

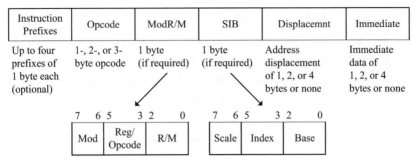

FIGURE 4.2: The x86 instruction format.

If we want to decode more than one instruction in parallel, we must know where each instruction starts. Having variable instruction length imposes a sequentiality on this task: we must know the length of instruction i before we can know where instruction $i + 1$ starts in the current fetch block. Thus, being able to quickly calculate the length of an instruction is critical for performance.

The first complication that an x86 decoder faces is identifying the instruction length. This is possible only after decoding the opcode and—if it exists—the ModR/M byte. The opcode defines whether there is a ModR/M byte and, if so, the ModR/M defines whether there is an SIB byte. The existence of displacement or immediate is also defined in the opcode.

There are two issues in decoding the opcode of an x86 instruction. First, the opcode is not always in the same offset from the beginning of the instructions. It could start anywhere in the first 5 bytes of the instructions, since we could have up to 4 bytes of prefixes before it. The second problem is that the opcode itself is of variable size—up to 3 bytes of primary opcode—and sometimes, bits 3 to 5 of ModR/M are used as an opcode extension.

The second complication that an x86 decoder faces is identifying the operands of the instruction. For example, in the simple case of a register-to-register operation, an operand can be encoded either in the opcode or in the ModR/M byte. The ModR/M byte, in turn, can encode 2- or 1-register operands, depending on the opcode and bits 6 to 7 of ModR/M.

In the register-to-register example, the 3-bit operand defines a general-purpose register, but to know which one, we need information from the opcode, the current execution mode and, in some cases, from the prefixes (if we have an operand-size override prefix). This is because with 3 bits, we can encode only 8 general-purpose registers, but there are many more architectural registers in x86. Thus, in 32-bit mode, an operand of value 0 can be interpreted as AL, AX, EAX, MM0, or XMM0 [19].

It is evident from the above discussion that x86 decoding is far from trivial. In modern x86 microprocessors, decoding takes several cycles, and it is a source of significant design complexity. In the following sections, we discuss possible decode unit implementations for high-performance, out-of-order, superscalar microprocessors.

4.3   DYNAMIC TRANSLATION

An x86 instruction has a lot of semantic information and may require several actions from the execution core. For example, the "add [eax], ebx" x86 instruction encodes an add operation of register EBX with the memory value at the address specified by EAX, with the result written back to memory at address EAX. This requires the processor to:

1. Calculate the address of the memory operand using EAX and the data segment register DS.
2. Bring the value of the memory location into the core and add to it the value of register EBX.
3. Store the result of the addition to the memory location calculated in step 1.

An out-of-order execution engine that tries to execute this instruction would require a lot of control state and signals to track at which stage of the instruction's execution it is at each point in time (this will be more aparent in Chapter 6). The microprocessor must guarantee that the stages of this instruction's execution happen in the correct order and that dependences with other instructions are guaranteed. If we want high performance, it would also be desirable to "parallelize" some parts of the execution of the instruction with other instructions. For example, the address calculation should not depend on a previous instruction producing the correct value for register EBX, but the addition stage should. Thus, it is evident that executing such complex instructions in an efficient manner on an out-of-order execution core is not an easy task.

On the other hand, a compiler for a RISC ISA would break this complex operation into three simple instructions. The following code sequence corresponds to such a code sequence (for a fictious RISC ISA). Here we use the x86 register names to make it easier to compare the two cases. Also, we assume the destination operand of the instructions is the leftmost one:

```
load r0, ds:[eax]
add r1, r0, ebx
store ds:[eax], r1
```

These instructions can be handled easily by an out-of-order execution engine. Each instruction will do one and only one operation, the dependences are clear for each instruction, the execution engine can intermix these operations with other nondependent operations freely, etc.

Because of the complexity the x86 CISC instructions introduce to the excution engine, early on, the x86 processor engineers decided to dynamically translate the x86 instruction stream to RISC-like instructions in the decode unit of the processor. The first implementations to follow this design were the AMD K5 and the Intel P6. This way, binary compatibility with the x86 ISA is maintained, while the execution engine—being similar to a RISC machine—is significantly simplified.

Nowadays, all modern out-of-order x86 microprocessors dynamically translate the x86 instructions into an internal RISC-like instruction format. In particular, Intel calls these internal instructions micro-operations, or μops for short. The P6 μops had a fixed length (118 bits) and a regular format, encoding an operation (i.e., the opcode) and three operands (two sources and a destination) [14]. The P6 μops use a load/store model. In the x86-vs.-RISC example above, the P6 decoder would generate a sequence of μops very much like the one of the RISC case.

From the size of the P6 μops, we can derive that actually, a μop corresponds not so much to a RISC instruction but more to a *decoded* RISC instruction, i.e., to the pipeline control signals of a simple RISC-like operation. Naturally, from their nature, the μops of modern microprocessors are different from the ones of the P6, but we believe that they still follow the RISC philosophy.

4.4   HIGH-PERFORMANCE X86 DECODING

Figure 4.3 shows the high-level block diagram of the decode unit of the Intel Nehalem architecture [21,23]. As it can be seen, x86 decoding is a multicycle operation. In this particular implementation, the process has been split into two decoupled phases: the intstruction length decoder (the "predecode" phase) and the dynamic translation to μops phase (the "decode" phase). The purpose of the ILD phase is to separate the raw byte stream into a sequence of valid x86 instructions and to pass these instructions to the second decode phase. The dynamic translation phase receives as input a stream of x86 instructions and generates a stream of functionally equivalent μops. The two phases are decoupled through the instruction queue (IQ). The reason for this—an alternative would be a simple latch—is to hide bubbles in the ILD that may appear when complex x86 encodings arise and to also allow the ILD to proceed when complex translations are required from the dynamic translation unit.

4.4.1   The Instruction Length Decoder

The ILD unit receives sixteen aligned bytes from the prefetch buffers and performs some basic predecoding to facilitate the dynamic translation phase. The ILD determines the length of each instruction, it decodes all its prefixes, and it also marks various properties of the instruction that will help the second-phase decoding [21]. Instruction length decoding is sequential by nature, so it must be as fast as possible if we want to be able to predecode many instructions at high frequency.

The most common instruction encodings can be handled in a single cycle by the ILD. In the Intel Core and Core 2 microarchitecture (and presumably in Nehalem as well), the following two cases cannot be handled by the normal path though, and a slow six-cycle path is used [21]:

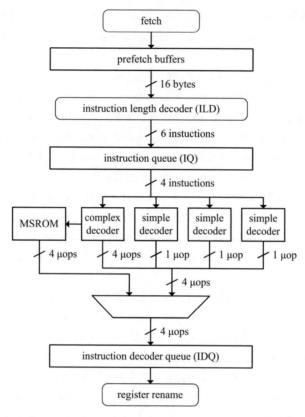

FIGURE 4.3: The Intel Nehalem decoding pipeline.

- An operand-size override prefix, preceding an instruction with a word immediate.
- An address-size override prefix, preceding an instruction with a ModR/M byte.

These two prefixes are termed length-changing prefixes (LCP). To determine the length of the instruction with an LCP, the ILD unit must do some complex decoding of the instruction including the operands, not just the opcode as in the normal ILD path.

4.4.2   The Dynamic Translation Unit

In this phase, instructions are read from the instruction queue, and they are translated to μops. Many of the x86 instructions, especially the register-to-register ones, are translated to a single μop. Some x86 instructions that have memory operands or that use complex addressing modes are translated to more than one μops.

The design of Figure 4.3 implements three "simple" decoders that handle instructions that are translated to a single μop and only one "complex" decoder that handles instructions that can be translated to up to four μops. This arrangement saves power and reduces complexity and, assuming that the greatest part of the instructions fall in the "simple" category, no decode bandwidth is sacrificed.

There are some x86 instructions, such as string instructions, that require more than four μops. These instructions are sent to the complex decoder, which then stops the normal decode pipeline and passes control to the microsequencer (MSROM) unit. The MSROM comprises a sequencer circuit and an ROM array. The microsequencer outputs a microcode program to emulate the complex x86 instruction. This program is nothing more than a preprogrammed sequence of normal μops (similar to those outputted by the other decoders).

· · · ·

CHAPTER 5

Allocation

Two main activities are performed at this pipeline phase: register renaming and instruction dispatch. The former has the purpose of removing false dependences due to reuse of registers, whereas the latter consists of reserving some resources that the instruction will require to be executed.

The dispatch consists of reserving some of the resources that instructions will use in the future, which normally include entries in the issue queue, the reorder buffer and the load/store queue. If any of the required resources is not available, the instruction is stalled until they become available. Sometimes, these resources are partitioned into multiple units, each one associated with some particular resources, so the allocation also has a side effect of determining which resources the instruction will later use for execution. For instance, there may be a different issue queue associated with each functional unit. In this case, the allocation also determines in which functional unit the instruction will be executed.

Register renaming is normally done only in out-of-order processors. Out-of-order processors execute the instructions of a program in an order that is different from the program order generated by the compiler or the programmer but has the same semantics. Instructions are reodered to increase the amount of instruction-level parallelism that is exploited. Instruction reordering is constrained by dependences among instructions. A dependence between two instructions forces a particular order between them. Dependences can be of two types (see Figure 5.1): data and name dependences. The former occurs when an instruction produces a data element that is consumed by another instruction. Obviously, in this case, the producer has to be executed before the consumer.

On the other hand, name dependences are caused by the reuse of storage locations, and they do not involve any particular data transfer among them. Name dependences can be of two types: write after write and write after read.

Name dependences are somewhat artificial; they are not due to the algorithm but to the fact that storage locations are reused. We could get rid of all name dependences by making every instruction write in a different storage location. This probably is not feasible for many applications/ systems since it would require more storage locations than available in today's systems (e.g., an application that runs for an hour in a single core may execute more than 10^{12} instructions). However,

r1 = r2 + r3	r1 = r2 + r3	r1 = r2 + r3
....	....	....
r4 = r1+r5	r1 = r4 + r5	r2 = r4 + r5
Data dependence	**Name dependence**	**Name dependence**
Read after write	**Write after Write**	**Write after read**

FIGURE 5.1: Instruction dependences.

even if there were enough memory, it may not be a good idea to do it, since it would cause a huge impact in performance due to the loss of locality.

Out-of-order processors have a much more modest target. They dynamically get rid of all name dependences but only for the in-flight instructions. Since typical instruction window size is around a hundred instructions, providing a different storage location for them is affordable. In this chapter, we are going to focus on register operands. Renaming is also applied to memory operands, as described in Chapter 6, when talking about the issue of memory instructions.

Most processors have a very small number of architectural registers (e.g., 32 integer and 32 FP) and, as a consequence, name dependences through registers are very common, and the benefits of getting rid of them in an out-of-order processor are huge.

Register renaming was first proposed by Tomasulo in his well-known scheme for out-of-order execution for the floating point unit of the IBM 360/91 in the 60s [43]. In that scheme, destination operands were renamed using the identifier of the reservation station that would produce them. This scheme is not used by current microprocessors, since it requires that the reservation station be occupied by an instruction until its execution completes. As described in Chapter 6, current microprocessors release the issue queue entries (reservation stations in Tomasulo's nomenclature) right after being issued, which is more effective in terms of efficiency.

There are three alternative renaming schemes that are used by contemporary microprocessors. We will refer to them as renaming through the reorder buffer, renaming through a rename buffer and merged register file.

5.1 RENAMING THROUGH THE REORDER BUFFER

In this scheme, register values are stored in the reorder buffer and the architectural register file. The reorder buffer (ROB) stores the results of noncommitted instructions, whereas the architectural register file stores the latest committed value for each architectural register. There is a rename table that indicates for every architectural register whether its latest definition is in the ROB or the architectural register file. In order to facilitate the access to operands in the ROB, the rename table

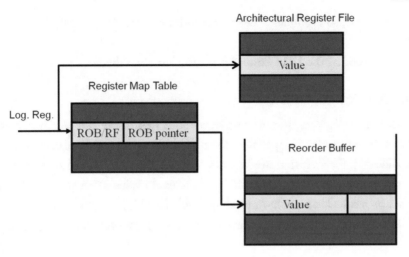

FIGURE 5.2: Renaming through the reorder buffer.

also contains an additional field in the former case that indicated the location in the ROB where the operand is (see Figure 5.2).

When an instruction executes, its value is stored in the ROB. When it later commits, the value is copied from the ROB to the architectural register file. Note that a given operand may reside on two different locations in its lifetime. This may introduce some extra complexity to the scheme to read operands, as discussed below.

This is scheme is used by some microprocessors such as the Intel Core 2.

5.2   RENAMING THROUGH A RENAME BUFFER

This scheme is a small variation of the previous one. The motivation is the fact that an important percentage of the executed instructions (around one third, although it varies a lot across applications and ISAs) does not produce any register result. In the previous rename scheme, each entry in the ROB has a field to store a register result, which implies that about one third of this storage is wasted. The idea of the rename buffer scheme is to have a separate structure for the result of in-flight (i.e., noncommitted) instructions. In this way, only instructions that produce a result consume a storage location. Like the reorder buffer scheme, results are first stored in the rename buffer and moved to the architectural register file when the instruction commits. When an instruction is renamed, if it requires a rename buffer entry and there is not any one available, the instruction is stalled until an entry becomes available (deadlock cannot happen since older instructions that are in flight cannot depend on the stalled one, so eventually, they will commit and free the rename buffer entries that they allocated).

This scheme is used by the IBM Power3 processor, among others.

5.3 MERGED REGISTER FILE

In this scheme, there is a single register file that stores both speculative and committed values. Because of that, the size of this file is bigger than the number of architectural registers. Each register is either free or allocated. Free registers are kept track of in a free list. Allocated registers are in use and may contain a committed value, a speculative value or no value at all (in the case that it has been allocated but its results have not been produced yet). In addition, there is a register map table that stores the latest assignment (physical register identifier) for each architectural register (see Figure 5.3).

The free list can be, for instance, implemented through a circular buffer that stores the identifiers of all free physical registers.

When an instruction is renamed, the rename map table is looked up to find out what its source operands are. In addition, if it produces a register result, a free physical register is allocated from the free list. If no free registers are available, the instruction is stalled until an older instruction commits and releases a register (see below for discussion on when registers are released). The destination operand is renamed to this free register, and the rename map table is updated to reflect this mapping.

Physical registers are freed when the processor can guarantee that they are dead (i.e., no instruction is going to use them anymore). Ideally, this could be done when the last instruction that uses this register commits. However, identifying the last use of a register is not straightforward for the processor (the compiler may know it, but current ISAs does not usually have a way to convey this information to the hardware). Because of that, the processor uses the following safe, conservative approach: a physical register is dead when the following instruction that uses the same archi-

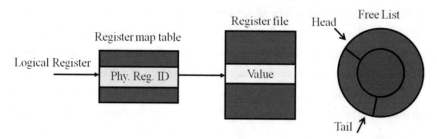

FIGURE 5.3: Merged register file.

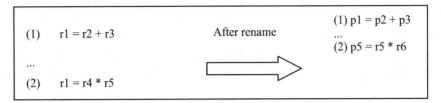

(1)   r1 = r2 + r3   After rename   (1) p1 = p2 + p3
                                                             ...
...   (2) p5 = r5 * r6

(2)   r1 = r4 * r5

FIGURE 5.4: Releasing a physical register.

tectural destination register commits. This is illustrated in Figure 5.4. In this example, register p1 is released when instruction (2) commits. Note that waiting until instruction (2) is fetched and all consumers of p1 before (2) commit is not enough, since instruction (2) may be squashed (e.g., due to a branch misprediction), and other consumers of p1 may be found later.

This is scheme is used by the Alpha 21264, MIPS R12000 and Pentium 4, processors among others.

5.4   REGISTER FILE READ

Another important aspect to consider is when register values are read, which has important implications in several key parts of the design. There are two alternatives: read before issue and read after issue.

In the former case, read before issue, the register file is read right before instructions are dispatched to the issue queue, and the values are stored in the issue queue. Obviously, not all the operands are available at that time, so only those available are read, and the rest are marked as nonavailable in the issue queue. Nonavailable operands later are obtained through the bypass network, and the register file is not accessed again. One advantage of this scheme is that the register file may require a fewer number of ports since only a portion of the operands is provided by it. On the other hand, the issue queue requires a storage for operands, which in some way is similar to a register file, and thus is expensive in terms of area. Besides, some source operands need to be read twice and written once: read from the register file and written to issue queue, and read from the issue queue to be sent to the execution units. This activity consumes energy, which is a very precious asset in processors nowadays.

In the latter case, read after issue, the issue queue stores the identifiers of the register source operands, and the operands are actually read after the instruction is issued to be executed. Operands that have just been produced and could not be written yet in the register file are obtained through the bypass network. This scheme requires a larger number of ports in the register file since there are more operands provided by it but, on the other hand, source operands are read just once and do not have to be copied anywhere (apart from the stage latches).

In theory, these two alternatives are orthogonal to the particular rename scheme being used, but there are important synergies that make some particular combinations quite common. In particular, for the rename schemes based on the reorder buffer and the rename buffer, read before issue has important advantages as described in the next section and is usually the preferred choice.

5.5   RECOVERY IN CASE OF MISSPECULATION

Instructions that are in flight have sometimes to be squashed due to multiple reasons (e.g., branch misprediction, exceptions). If these instructions have gone through the allocate stage, then the resources that they reserved have to be released. Besides, the modifications that these instructions did in the rename tables have to be undone so that they reflect the same state they would have if these instructions never would have been executed. Chapter 8 discusses how this recovery is performed.

5.6   COMPARISON OF THE THREE SCHEMES

Regarding the allocation and release of physical registers, the scheme based on the ROB is very simple. There is no need to keep a free list since physical registers are part of the ROB entry assignment, which is basically a FIFO structure. For every new instruction, the tail of the FIFO is allocated, and when an instruction commits, the head of the FIFO is released. The same is true for the rename buffer scheme, since the rename buffer also is managed as a FIFO structure. On the other hand, the merged register file has a more complex management scheme. It requires a free list of physical registers. When a new physical register is needed, the head of the list is used. On the other hand, to release physical registers, each instruction has to keep in the reorder buffer the identifier of the physical register that was mapped to the destination register right before this instruction was renamed. For instance, in the sample code of Figure 11, instruction (2) will store in the reorder buffer the identifier of p1, so that is released when (2) commits.

On the other hand, the merged register file has two main advantages. First, register values are written just once and never move, whereas for the other two schemes, they are written twice, first in the reorder buffer or rename buffer and later in the architectural register file. This extra activity represents additional energy consumption. Second, in the merged register file, all source operands come from a single location, whereas in the other two schemes, they may come from two different locations (the architectural register file or the reorder buffer/rename buffer). Having a single location reduces the amount of interconnect that is needed and potentially may be beneficial in terms of area spent by the interconnect between the register file and the functional units.

Finally, the merged register file scheme can be used with the two read approaches described above (read before issue and read after issue) with no significant differences regarding its implementation. On the other hand, the reorder buffer and rename buffer schemes are more appropriate for read before issue and present some challenges if one wants to use them together with read after issue. In particular, the challenge comes from the fact that the register values eventually move from one location (reorder buffer or rename buffer) to another (architectural register file). In the read-after-issue scheme, the issue queue stores the identifier of the source operands. If when an instruction is renamed, a source operand is in the reorder buffer or the rename buffer, the issue queue will store a pointer to that location. If the instruction that produces this source operand commits before the operand is read by the consumer, the value will be moved to the architectural register file, and the pointer stored in the issue queue will not be correct anymore, since this entry may be allocated by a different instruction. In order to correct it, it would be necessary to do an associative search in the issue queue for every committed instruction to check if any entry is pointing to its destination register. If this is the case, the pointer should be changed to the corresponding architectural register file entry. All of this is very complex in hardware. The associative search is similar to the wakeup logic described later and, on top of that, additional write ports would be required to store the new pointer. Because of this, processors that use renaming through the reorder buffer or through the rename buffer normally opt for the read-before-issue scheme.

. . . .

CHAPTER 6

The Issue Stage

6.1 INTRODUCTION

The issue is the pipeline stage in charge of issuing instructions to the functional units for execution. There are two main types of issue schemes: in order and out of order. The former one steers the instructions in program order, whereas the latter steers instructions out of order as soon as their source operands become available.

In general, in-order schemes check the oldest nonissued instruction and issue it whenever its source operands and the resources needed for its execution are available [46].

However, most of the latest processors implement out-of-order schemes. There are many different ways of implementing an out-of-order issue. Indeed, the final implementation is also very dependent on the design decisions made for the rest of the stages. For instance, we do not need the same hardware if the source operands of the instructions are read before or after the dispatch stage. Also, there are significant differences among issue schemes depending on whether they are based on reservation stations [30], distributed issue queues [27,33,48], or unified issue queues [39].

In this chapter, we describe the most common issue schemes implemented in existing processors.

6.2 IN-ORDER ISSUE LOGIC

In-order issue logic issues the instructions for execution in the same order they were fetched. Therefore, instructions wait until all previous instructions have been issued. Then, the instruction is issued as soon as its source operands are available and the resources it needs for execution are ready.

This kind of issue logic is sometimes implemented at the decode stage of the processor due to its simplicity using scoreboarding. A typical scoreboard comprises two tables, a data dependence table and a resource table. These tables may vary depending on the actual hardware constraints of the processor.

The data dependence table is indexed using the source register identifiers of the instruction to be issued. Every entry on this table represents the state of a register value. This state ranges from nonavailable, so that the instruction cannot be issued yet, until value available, either at some bypass level or written into the register file. The reader will find more details regarding bypasses in Chapter 7.

The resource table keeps track of the availability of execution resources like functional units. There are some functional units like divisors that are not able to accept one new operation request every cycle. In this case, the processor could not schedule an instruction that uses the divisor if it scheduled another instruction that used it one cycle before. Therefore, the issue logic uses this table in order to check whether a given execution resource is available on the current cycle.

Very long instruction word (VLIW) processors implement a simplified in-order issue logic. These processors do not implement any kind of scoreboarding since it is the responsibility of the software that generates the code to schedule every instruction far enough from the producer to have its inputs available when it is issued for execution. This software is usually a static compiler or a codesigned virtual machine like in Transmeta Efficeon [47].

6.3   OUT-OF-ORDER ISSUE LOGIC

The issue logic is a key component that determines the amount of instruction-level parallelism out-of-order processors are able to exploit. It allows out-of-order execution by issuing instructions to the functional units as soon as their source operands become available. However, the hardware components involved in the issue process sit in the critical path of the processor pipeline [1]. Therefore, it is very important to implement a good complexity-effective issue logic able to exploit instruction-level parallelism without compromising the cycle time.

There are many different alternatives to address the multiple design decisions involving the implementation of an issue logic. However, the goal of this chapter is not to give a wide description of all possible implementations but to show the most common examples with the aim of giving an idea of the characteristics of the hardware.

In this chapter, we cover two main scenarios assuming a unified issue queue. Processors that use a unified issue queue implement a single queue where all renamed instructions are stored, waiting to be executed. This is different from other schemes like reservation stations or distributed issue queues where instructions are allocated in separate buffers depending on the type of resources they need for its execution.

The first scenario represents an implementation of the issue logic for processors where instructions read their source operands before entering the issue queue like P6-like architectures. Then, as second scenario, we describe the main changes required to implement the issue logic where source operands are read after they are issued for execution like in MIPS R10000 or Alpha 21264. These two scenarios are suitable for any of the different existing schemes to hold the values produced by the instructions (merged register file, rename buffers, reorder buffer, etc).

Nevertheless, since this is an orthogonal design decision, for the sake of clarity, we will assume a merged register file for both implementations. Note that we call a merged register file to a register file that stores the architectural state and the speculative values as described in detail in Chapter 5. However, the described hardware easily can be adapted to any other register file scheme.

This chapter also covers other alternatives like distributed issue queues and reservation stations. These alternatives will be explained in less detail since most of the tradeoffs that need to be considered in the implementation already have been covered with the aforementioned scenarios.

Finally, we pay special attention to the implementation of the issue logic for memory operations. Conversely to the rest of operations where data dependences are checked at the renaming stage, memory dependences cannot be identified until the memory operations compute their addresses. This characteristic has significant implications on the management of these instructions, as we will decribe later.

6.3.1 Issue Process when Source Operands Are Read before Issue

The main characteristic of an issue queue where operands are read before the issue stage is that it needs to hold the information from the instruction to perform the issue and the values from the source operands that have been already produced. Figure 6.1 shows a general overview of the typical components used to store this information. Every block in Figure 6.1 represents a table with as many entries as the number of instructions that can be held by the issue queue. Moreover, for the sake of simplicity, we assume a processor with an ISA similar to a simplified MIPS [32], where instructions can have up to two source operands or one source operand and an immediate value coded as part of the instruction.

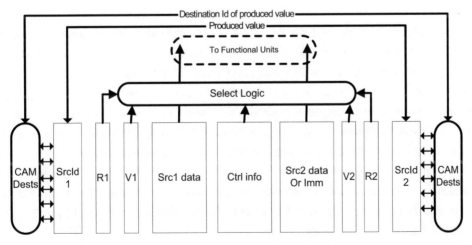

FIGURE 6.1: Hardware components of a typical issue queue where source operands are read before issue.

In our simple example, there is a block called *Ctrl info*. This block is in charge of holding all the static information (type of required ALU, data size for memory operations, use of immediate operand, etc.) needed for the execution of the instruction. Then, there are two symmetric arrays called *Src1Id* and *Src2Id*. These arrays store the source operand identifier for source 1 and source 2, respectively. These identifiers are unique in the processor and have been generated by the renaming stage, as described in Chapter 5. In case one of the source identifiers is not used because the instruction does not have two source operands, the source ID for that operand becomes invalid. Invalid source operands are identified by using the valid arrays *V1* and *V2*. These blocks implement 1 bit per entry that says whether *Src1Id* and *Src2Id* are valid, respectively. In this example, we assume that immediate values always go to *Src2*. Therefore, in case the instruction has an immediate value, the valid array would mark the source as valid, but the *Src2Id* would be set to 0 or any other identifier that is never used for renaming purposes. 0 is a good alternative since register 0 usually is hardcoded to 0 so that it is never a useful destination.

Blocks *Src1 data* and *Src2 data or Imm* are in charge of storing the input values. If the instruction has an immediate, it is stored in block *Src2 data or Imm*.

Finally, the ready bit arrays *R1* and *R2* notify whether the *Src1 data* and *Src2 data or Imm* have their values already produced. As soon as the ready bits are set for all the valid sources of an instruction, the instruction can be steered for execution.

Once we have introduced the main components, we describe the different events that occur on the issue queue and how this hardware structure interacts with the rest of the processor. These events are issue queue allocation, instruction wakeup, instruction selection and issue queue reclamation.

6.3.1.1 Issue Queue Allocation. Figure 6.2 shows an example of an integration of the issue queue inside a generic pipeline of an out-of-order processor. As we can see, the instructions are first renamed at the renaming stage (a.k.a. allocation stage) and allocate and enter in the issue queue. In case there are no available entries, the allocation stage is stalled. Note that the renaming stage may not be able to use up-to-date information about the issue queue occupancy due to time constraints. In this case, the renaming could make conservative assumptions about the issue queue occupancy to avoid processing instructions that may not find issue queue entries available.

The register file is then accessed during the next cycle in order to read the source operands that are already available. Note that the renaming table does not only store the register ID for each source operand but also an *available* bit that says whether this is already available. Therefore, an instruction will have to read from the register file those inputs whose *available* bit is set.

Finally, the issue queue entry associated with the renamed instructions is filled with the renaming information plus the data read from the register file.

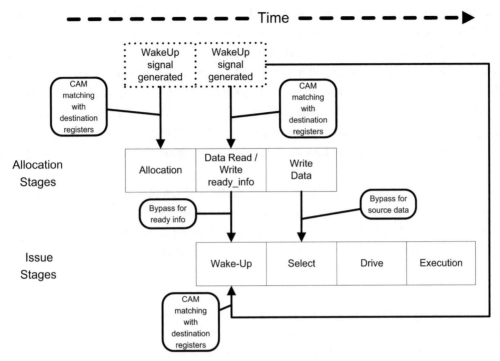

FIGURE 6.2: Example of the pipeline stages involved in the instruction issue on a generic out-of-order pipeline.

6.3.1.2 Instruction Wakeup. Wakeup is the event that notifies that one of the source operands has been produced. This signal usually comprises the renaming ID of the produced value, the value itself and a valid bit. Then, the CAM logic compares this ID with the entries in the *SrcId1* and *SrcId2* arrays, as shown in Figure 6.1. In case of a match, the corresponding ready bit is set and the value copied on the corresponding *Src Data* entry. As soon as the ready bits of the valid sources of an instruction are set, the instruction becomes ready (we say it is woken up) and can be considered by the select logic to be steered for execution.

Note that the wakeup signal is produced only once, and consumer instructions may not be in the issue queue when it occurs. Therefore, it should be guaranteed that once a value is produced, all its consumers will know that the data is already available. This could be done by setting to 1 the aforementioned *available* bit at the renaming table.

Processors that read operands before issue usually require at least one extra cycle between the renaming stage and the allocation of instructions into the issue queue, as shown in Figure 6.2. Note that source operands of the instructions at this stage should also be compared with the coming wakeup IDs to prevent deadlocks. Otherwise, if some of their source operands are produced in this cycle, these instructions would never wakeup, since they read the *available* bit the cycle before, and they do not write the issue entry to perform the CAM matching until the next cycle.

A possible solution that would reduce the amount of pipeline stages that require CAM matching is to write *SrcId1* and *SrcId2* in advance. This information could be written right after it is obtained from the renaming table. Figure 6.2 shows a scenario where this information is written the cycle after renaming. Then, the issue queue will do the conventional comparisons between the wakeup events and the *SrcId* fields in order to keep the ready bits up-to-date as it happens with the rest of the entries. Once the values have been read from the register file, the *Src1 data* and *Src2 data* will be updated accordingly.

Another alternative that would save us from implementing CAM matching logic for all stages between renaming and the issue queue allocation is to implement the *available* bit as a separate table from the renaming table. In this case, the renaming table would be accessed at the renaming stage, but the *available* bit table would not be accessed until the instruction allocates their *SrcIds* into the issue queue.

Advancing *SrcId* allocation has some advantages over the other alternative since it allows an instruction to wake up at the same cycle it reads the source operands. Thus, it could be considered for execution the same cycle it writes the whole data into the issue queue, as shown in Figure 6.2.

As has been commented before, the wakeup signal notifies that a value is already available. However, this signal may be generated before the value actually is produced in order to minimize the distance between the execution of the producer and the consumer. The wakeup signal can be notified in advanced because an instruction does not need its source operands until it reaches the execution stage, one or more cycles after the instruction wakeup. For instance, the pipeline in Figure 6.3 implements two stages between wakeup and execution.

Figure 6.3 shows the number of cycles between the execution of the producer and the consumer depending on the moment when the wakeup signal is sent. In scenario 1, the wakeup signal is sent as soon as the producer is executed and the value is available. Then, the consumer is woken up the next cycle and executed three cycles after the producer generated the value. This three-cycle bubble could be avoided by generating the wakeup signal earlier, as shown in the scenario 2. In this case, the wakeup signal is sent three cycles before the producer generates the value. Then, the consumer will reach the execution stage right after the producer generates the value, allowing back-to-back execution. Note that in the second scenario, the value should be transferred from the output of the functional unit to the input of another functional unit. This connection is called bypass, and its implementation is described in Chapter 7. Moreover, the scenarios shown in Figure 6.3 also assume that the select of the producer and the wakeup of the consumer can be done in a single cycle. This design decision is also critical for performance because if the select and wakeup do not fit in one cycle, back-to-back execution cannot be performed for instructions with one cycle latency

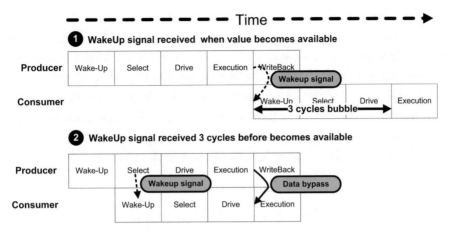

FIGURE 6.3: Timing of the wakeup signal to support back-to-back execution.

incurring in significant performance drops [7]. In conclusion, back-to-back execution is critical for performance [34], and for this reason, most of the processors implement it [27,30,33,48] among others.

There are two common implementations to generate the wakeup signal. One alternative is to generate the signal in the pipeline stage where the instruction resides three cycles before its execution completes. Note that the number of cycles an instruction requires for execution depends on the functional unit it uses. For instance, an integer adder usually requires one single cycle to complete, whereas an integer multiplier or a floating point functional unit may require longer. Therefore, the pipeline should be able to generate the wakeup signal from the select stage, for single-cycle operations, until three cycles before the functional unit that takes longer ends.

Another alternative is to implement every entry of the valid bit array as a shift register plus the valid bit. These shift registers also may be implemented as a scoreboard with one shift register per physical register. Every shift register should have as many bits as the maximum number of cycles required by a functional unit to produce the value. Then, the wakeup signal is always generated at the select stage, and it sets to 1 the bit of the shift register in the position equal to the latency of the functional unit minus 1. The shift registers shift every cycle and, as soon as the bit 0 of the shift register becomes 1, the corresponding valid bit is set.

Note that these mechanisms are suitable when the latency of an instruction is constant and only depends on the instruction itself. This assumption applies to all arithmetic operations but not for memory operations. The latency of a memory operation (a load, for instance) depends on whether it hits or misses in the data cache or the data TLB. Unfortunately, this is known only when the load is issued, computes its address and accesses these structures.

Loads could be handled in a conservative manner by delaying the generation of the wakeup signal until we know whether the load will hit in the cache and the TLB. Then, in case of a, hit we could immediately wakeup the consumers. In case of a miss, we would not generate the wakeup signal until the miss is solved. However, the average number of load operations in a program is around 20% [16], and most of them have consumers, so that delaying the wakeup signal for these operations would have significant impact on performance. Therefore, some processors speculatively wakeup load consumers assuming hit latency and pay a penalty in the infrequent cases when a load misses in cache. The speculative wakeup is explained later in this chapter.

6.3.1.3 Instruction Selection. The selection logic (a.k.a. select logic) is in charge of choosing the subset of ready instructions in the issue queue that will be steered in a given cycle.

An instruction can be selected if its source operands are ready and the execution resources it requires are available. For instance, it is not possible to steer two multiply instructions in parallel if the processor does implement only one multiplier.

The timing of the selection logic is very critical since it has to be done after the wakeup logic to support back-to-back execution of single-cycle latency operations. For this reason, processors do not usually implement a single selection logic, but they distribute it into simple components called arbiters or schedulers [27,33,39]. For instance, a processor able to steer up to 4 instructions per cycle would implement either 2 or 4 arbiters, and every functional unit will be statically bound to a single arbiter. Instructions waiting on the issue queue are also bound to a single arbiter. This configuration allows parallelizing the selection logic for every functional unit. Otherwise, the arbiters should be synchronized to guarantee that they do not select the same instruction or they do not steer two different instructions to the same execution resource.

Figure 6.4 shows a possible implementation of a selection logic based on arbiters, as in the Alpha 21264 [33]. In this simplified example where we only focus on integer arithmetic operations, the maximum issue width is four, but limited to a maximum of three simple arithmetic operations and one multiply. Note that 4-wide issue processors do not usually allow issuing 4 instructions of any kind, but they are constrained by the functional units implemented. Common processor designs implement several units that execute simple arithmetic operations but only one or few of them that support complex arithmetic operations. The example shown in Figure 6.4 splits the issue width into two arbiters *A* and *B*, where arbiter *A* is bound to one simple funcitonal unit and a multiplier, whereas arbiter *B* is bound to two simple functional units.

Instructions in the issue queue are also bound to an arbiter. Once an instruction is renamed, it is assigned to an arbiter by the arbiter steering logic and allocates an entry on the subarray of the specified arbiter. This steering logic is usually very simple and sends instructions to the arbiter,

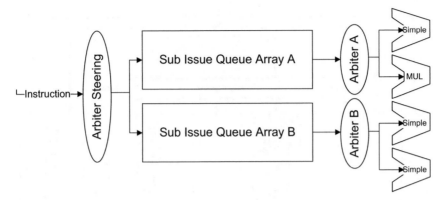

FIGURE 6.4: Implementation of a selection logic based on arbiters.

whose functional units could execute them, trying to keep the same amount of instructions allocated per arbiter.

Every cycle, each arbiter checks its subarray for ready instructions available and selects the ones to be steered for execution. In the case of the Alpha 21264, when the number of ready instructions in the subarray exceeds the issue width, the oldest instructions always have priority. This algorithm is relatively easy to implement in processors like the Alpha 21264, where instructions always are stored in order in the issue queue subarrays. However, other processor designs like MIPS or Intel P4, where the instruction order is not preserved in the issue queues, either implement pseudo-age-based algorithms or simply prioritize based on the position inside the issue queue [48].

6.3.1.4 Entry Reclamation. Once an instruction has been selected and its data forwarded to the functional units, its issue queue entry can be safely reclaimed. However, some techniques like speculative wakeup may require delaying the reclamation until we are sure the instruction can be executed. Speculative wakeup typically is used to reduce the latency between a load operation and its consumer. This technique is covered in Section 5.2.4.

6.3.2 Issue Process when Source Operands Are Read after Issue

This section describes the main differences between a scheme where source operands are read after the instruction is issued and the previous scenario.

The key difference between reading after issue and reading before issue is that reading after issue does not require the issue queue to store the source values. Therefore, the wakeup signal does not need to forward the values. Figure 6.5 shows the difference between the components of an issue queue when data is read before issue and after issue. The gray components are those that

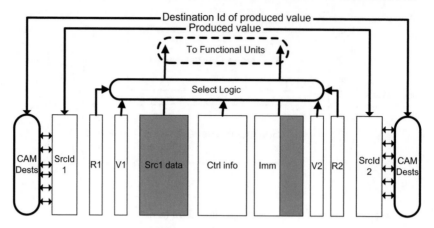

FIGURE 6.5: Issue queue components when source operands are read after issue compared to an issue queue when data is read before issue. The grayed components are those required by the latter scheme that are not present in the former one.

are not required if we read after issue. Note that even though we fully remove *Src1 data*, *Src2 data* still remains, although it could be smaller. The reason is that this scheme still needs room to hold immediate values. However, these values are usually shorter than register values. Besides, there are few instructions that require this field so that it could be implemented as a separate structure with few entries. In this latter implementation, the issue queue entries that require an immediate would include the offset inside this structure where the immediate resides.

Figure 6.6 shows a possible pipeline for this issue scheme. As can be seen, the number of stages between the renaming stage and the issue queue allocation is reduced since data is not read yet. This implies a reduction in the number of required CAM-matching logic.

However, the number of cycles between the wakeup and the execution stages is increased by one cycle in order to read the source operands.

Another significant difference between reading before and after issue comes from the number of read ports required in the register file. Whereas the number of read ports required in the register file when reading before issue is determined by the machine width, which typically ranges between 3 [27,39] and 4 [48], the number of read ports when reading after issue is determined by the issue width. Although it may look counterintuitive, the issue width is sometimes wider than the machine width. The reason is that the issue arbiters are not generic, but they are specialized in certain types of instructions. For instance, an architecture with specialized arbiters for arithmetic and memory operations able to execute up to 4 integer arithmetic operations or 4 memory operations would require an issue width of 8. We can find an example on the Netburst architecture [27] that is a 3-wide machine able to issue up to 6 instructions per cycle.

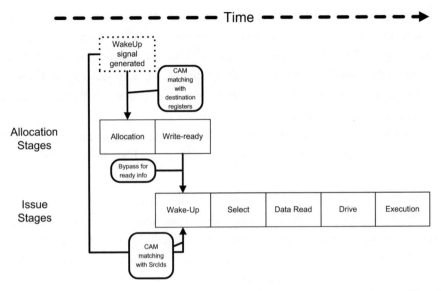

FIGURE 6.6: Example of the pipeline stages involved in the instruction issue when reading after issue.

6.3.2.1 Read Port Reduction. The area, power and access latency of the register file increases with the number of read ports [7]. Therefore, it is important to minimize the number of ports in order to achieve a power-efficient design.

Some processor designs implement as many read ports as needed assuming the worst case scenario where the issue width is fully utilized and all instructions read their operands from the register file [29,33,48]. However, the Alpha 21264 reduces the number of read ports per register file by splitting it into two replicated physical register files with half the total number of read ports each. This solution reduces the number of read ports at the expense of penalizing in one cycle the back-to-back execution between producers and consumers that access different register files.

It has been shown in the literature [7] that most of the sources are read from the bypass network instead of from the register file. Moreover, the issue width usually underutilized. Therefore, it also may be possible to implement a register file with fewer ports than needed in the worst case scenario with minimal impact on performance.

There are two possible alternatives to do this: active and reactive. The active alternative consists of synchronizing the arbiters in order to compute the number of read ports that will be used by each of the selected instructions. If the number of required read ports exceeds the available ones, some of the arbiters will cancel the issue process. Note that the reason why the issue logic is distributed in arbiters is to reduce the latency of this logic by fully parallelizing it. Therefore, implementing this read port's synchronization may have impact on the delay.

The reactive alternative makes the arbiters issue instructions assuming that the total amount of read ports will never be reached and reacts in the rare case when it occurs. In this alternative, read ports are assigned when instructions are issued. Then, if the number of available read ports is exceeded, some of the issued instructions should be cancelled and reissued again. Cancellation and reissue could be done using any alternative implemented in processors that perform speculative wakeup. Some of these techniques are described in 5.3.

Finally, note that reactive mechanisms may incur in starvation among arbiters and even live locks. Thus, it is important to define a fair policy in order to perform the cancellation.

6.3.3   Other Implementations for Out-of-Order Issue

The discussion of the previous scenarios covers most of the main design decisions that need to be addressed on the implementation of an issue logic. However, for the sake of completeness, we briefly describe in this section two other issue implementations that can be found in modern processors: distributed issue queue and reservation stations.

6.3.3.1   Distributed Issue Queue. Processors that implement this scheme distribute the functional units in exection clusters where each of the clusters implements its own issue queue. For instance, Intel Pentium 4 implements two execution clusters with private issue queues: one for memory operations and another for nonmemory operations. In this case, instructions are steered to one of the issue queues depending on operation type, and then it is bound to a specific scheduler inside the issue queue.

6.3.3.2   Reservation Stations. Reservation stations are private buffers per functional unit that store the instructions that are going to be executed on the specific functional unit and their input values. This scheme was proposed by R. Tomasulo for the IBM 360/91 in 1967, and it is the basis of modern superscalar processors.

Processors implementing this scheme steer instructions after renaming to the buffer of the specific functional unit where they wait until their input values become valid. Then, every instruction broadcasts its produced value to all reservation stations of all functional units. As can be seen, this technique requires some design decisions similar to the issue queue scenario where data are read before issue.

6.4   ISSUE LOGIC FOR MEMORY OPERATIONS

Memory operations have data dependences through memory that cannot be identified at the renaming stage. These memory dependences can be checked only once these instructions have been issued and their addresses computed.

The mechanism in charge of handling memory dependences is called memory disambiguation policy. Different processors implement completely different memory disambiguation policies. Table 6.1 shows some typical schemes implemented on different microarchitectures. These schemes can be classified into two main groups: nonspeculative and speculative disambiguation policies. The first group does not allow executing a memory operation until we are sure it does not have dependences with any previous memory operation. By contrast, the second group predicts whether a memory operation will have a dependence with another in-flight memory operation.

The selection of a proper memory disambiguation is critical for the performance and complexity of a processor design. Around 30% of the instructions executed by a processor are memory operations. Therefore, implementing very conservative memory disambiguation policies may produce an unnecesary serialization of the execution that could significantly limit the instruction-level parallelism that can be exploited. On the other hand, very aggressive memory disambiguation policies may end up on complex recovery mechanisms and significant power increase due to misspeculations.

6.4.1   Nonspeculative Memory Disambiguation

Nonspeculative memory disambiguation policies do not issue any memory operation until all previous stores have computed their addresses. Therefore, memory dependences safely can be computed. There are three main types of nonspeculative memory disambiguation policies implemented by existing processors: total ordering, load ordering with store ordering and partial ordering.

TABLE 6.1: Memory disambiguation schemes.		
NAME	**SPECULATIVE**	**DESCRIPTION**
Total Ordering	No	All memory accesses are processed in order.
Partial Ordering	No	All stores are processed in order, but loads execute out of order as long as all previous stores have computed their address.
Load Ordering Store Ordering	No	Execution between loads and stores is out of order, but all loads execute in order among them, and all stores execute in order among them.
Store Ordering	Yes	Stores execute in order, but loads execute completely out of order.

In total ordering, all memory operations are executed in order. Nowadays and from the best of our knowledge, there are no out-of-order processors that implement total ordering, because it constrains a lot the amount of parallelism we can exploit.

By contrast, the rest of the nonspeculative schemes allow load operations to execute out of order with respect to stores. In the case of load ordering with store-ordering schemes, loads proceed in order, and stores proceed in order. However, loads do not have to wait for previous stores to access the cache. We can find this scheme implemented in processors like the AMD K6. In partial ordering, though, loads can be processed out of order. In this case, a load can be issued as long as it has its source operands ready and all previous stores already have computed its address. Examples of processors implementing partial ordering are the MIPS R10000 and the AMD K8.

Note that the memory disambiguation can be performed as soon as the memory addresses of the stores are computed. Thus, some processors split the store operations into two subtasks: one that computes the address and another that gets the data. Then, the store operation does not have to wait for the producer of the data to complete in order to compute its address. In some cases, even the processor computes the address as soon as possible and does not read the data until the store becomes the oldest in-flight instruction, like in the HP PA8000 [29].

In the next sections, we present two case studies of nonspeculative memory disambiguation policies implemented in existing processors. The first example is the AMD K6 processor that implements the load-ordering, store-ordering scheme. The second example shows the implementation of partial ordering on a MIPS R10000.

6.4.1.1 Case Study 1: Load Ordering and Store Ordering on an AMD K6 Processor. The AMD K6 processor implements two separate pipelines for load and store operations. Both pipelines are decoupled with some level of communication, but instructions flow in strict order inside each pipeline. A simplified example of these two pipelines can be found in Figure 6.7.

This memory pipeline implements the following components in order to perform the disambiguation:

- **Load queue:** this queue stores the load operations in program order. Loads are inserted in this queue after renaming and reside there until they become the oldest on the queue and their source operands are ready.
- **Address generation:** this is the logic in charge of computing the access address of a memory operation based on its source operands.
- **Store queue:** this queue stores the store operations in program order. Store instructions reside here since they have been renamed until they become the oldest instruction on the queue; the source operands they need to compute the address are available.

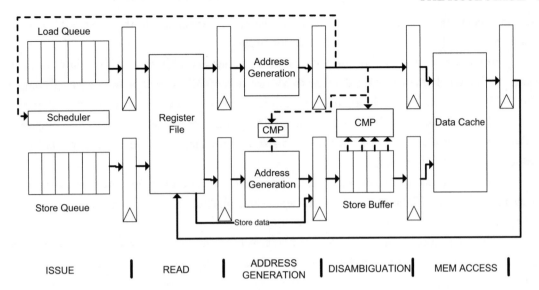

FIGURE 6.7: Schematics of the AMD K6 pipeline to implement a load-ordering store-ordering memory disambiguation policy.

- **Store buffer:** this buffer keeps the store operations in program order until they become the oldest in-flight instruction in the processor, and then they proceed to update the memory.

Loads flow through the upper part of the pipeline, as shown in Figure 6.7, whereas stores flow through the pipeline at the bottom. These operations are processed on every pipeline stage as follows.

Loads as well as stores are issued on the issue stage as soon as their source operands are ready and they become the oldest instruction on the load and store queue, respectively. Note that in the case of a store, the issue logic does not wait for the data to be stored in memory to be ready but only for the source operands required to compute the store address.

Then, loads and stores read the source operands required to compute the address from the register file on the read stage. These data may be either read from the register file or obtained from the bypass logic.

On the address generation stage, the address the memory operation will access is computed. In case of a store, the value to be stored in memory is read. If this value is not available, the whole pipeline for store operations is stalled.

Once the address is computed, the operations move to the disambiguation stage. In the case of a store, both the data and the address are stored in the store buffer. This information will reside in the store buffer until the operation becomes the oldest instruction in the processor. Then, the cache will be updated accordingly on the memory stage.

In the case of a load, the load will compare its memory address with the addresses of the stores in the store buffer that are older than it. Moreover, the load also will perform a partial comparison with the store that is on the address generation stage in case this store is older. This comparison is partial because this store does not have time to fully compute its address before the comparison takes place. Therefore, only few bits are compared, and the load is considered dependent on this store in case these bits match.

Finally, the load checks the scheduler to be sure that there is no older store that has not computed its address yet. Then, the load and the whole load pipeline are stalled if the load hits with any previous store or there are older stores on the issue queue.

We may think that processing loads in order is an unnecessary constraint since loads do not modify the memory, and they do not have dependences among them. However, this is a simple way of implementing processor consistency as specified on the x86 reference manual: stores have to be visible in order, and loads have to be perceived as executed in order. It is possible, though, to support processor consistency without serializing the execution of the loads. For instance, AMD improved its memory disambiguation scheme on the AMD K8L by allowing loads to overtake previous loads under certain circumstances and still preserving the support for processor consistency. Another example is the Intel Core architecture. In this case, processor consistency also is supported on top of a speculative memory disambiguation model that executes loads out of order, even overtaking stores that have not computed their addresses. The description of mechanisms to guarantee processor consistency in processors that execute memory operations out of order is out of the scope of this chapter.

6.4.1.2 Case Study 2: Partial Ordering on a MIPS R10000 Processor. The MIPS R10000 processor implements partial ordering. Therefore, loads can be executed out of order as long as all previous memory operations have computed their addresses. By contrast, stores are processed in strict program order. The schematics of the pipeline stages and components involved in the memory disambiguation are shown in Figure 6.8.

This memory pipeline implements the following components in order to perform the disambiguation:

- **Load/store queue:** this is a 16-entry queue where loads and store instructions are stored in order after the renaming stage. Instructions do not leave this queue until their source operands are ready.
- **Indetermination matrix:** this is a 16x16 half matrix where every column and row represents an entry on the load/store queue. A memory operation sets all entries on its column

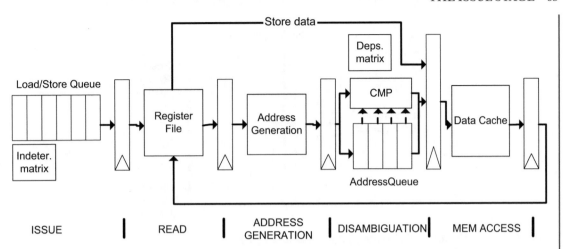

FIGURE 6.8: Schematics of the MIPS R10000 pipeline to implement the partial ordering memory disambiguation policy.

to 1 and resets them when it computes its address. Then, a memory operation cannot be issued while there is a 1 on any position of its row belonging to an older memory operation. Figure 6.9 shows an example of an indetermination matrix for a 6-entry load/store queue.

- **Dependency matrix:** this is a 16x16 matrix where every column and row represents an entry on the load/store queue. A load that depends on previous stores set to 1 all entries on its row for all the columns belonging to the stores it depends on. Then, it will not be able to resume its execution until all entries on its row are reset. By contrast, a store resets all bits on its column when it updates the memory. Figure 6.10 shows an example of a dependency matrix for a 6-entry load/store queue.
- **Address generation:** this is the logic in charge of computing the access address of a memory operation based on its source operands.
- **Address queue:** this queue keeps the memory addresses of loads and stores that want to access the cache. In the case of loads, besides writing their address on this queue, they also compare it with the addresses of all previous stores and, in case of matching, the corresponding entries on the dependency matrix are set.

All memory operations activate the bits of their column on the indetermination matrix at the renaming stage. Note that the column associated with the instruction depends on the position it occupies on the load/store queue. Then, as soon as its row on the indetermination matrix does not include any one (all previous memory instructions have computed its address) and its source operands are available, the memory operation is issued and read its source operands at the read stage.

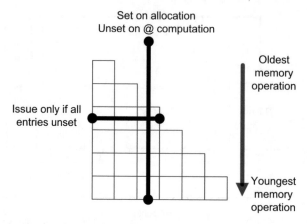

FIGURE 6.9: Example of a 6-entry indetermination matrix.

Once the memory operation has computed its address, it is stored on the address queue. In the case of a store, it will reside there until it becomes the oldest instruction in the pipeline. Then, it will be issued resetting its column on the dependency matrix and read the data to be stored in memory from the register file using a dedicated read port. Note that these data are already computed because since the store is the oldest instruction in the pipeline, all previous instructions have produced their outcome and retired already. By contrast, loads will compare their address with the addresses of previous stores and update their row on the dependency matrix accordingly. A load will wait on the address queue until its row on this matrix is fully reset.

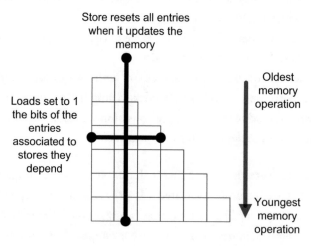

FIGURE 6.10: Example of a 6-entry dependency matrix.

The MIPS R10000 implements a 2-way data cache where instructions access out of order. In case of a cache miss, the instruction has to wait until the miss is solved, but other memory instructions can proceed in the meantime. This situation may incur on an undesirable situation called thrashing. Thrashing occurs when there is a memory operation miss on cache, and by the time the miss is solved and before it reads the data, another instruction has evicted the line again. Note that thrashing may incur on livelocks if it affects the oldest instruction in the pipeline. MIPS R10000 avoids trashing by locking a way of the set that will be accessed by the oldest memory operation in the pipeline until it successfully reads the data. The oldest instruction locks a way of its set on the disambiguation stage.

Finally, loads and stores access memory as soon as they are allowed to leave the address queue, as commented before.

6.4.2 Speculative Memory Disambiguation

Some of the latest processor designs like the Alpha 21264 or the Intel Core architecture already implement speculative memory disambiguation. These processors boost performance by speculatively issuing loads that are predicted not to be dependent on any previous in-flight store. Therefore,

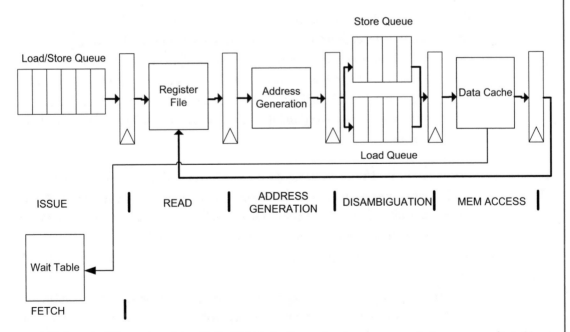

FIGURE 6.11: Schematics of the Alpha 21264 pipeline to implement a speculative memory disambiguation policy.

load operations do not have to wait for all previous stores to compute its address. Note that since this scheme speculates on the memory dependences, it may happen that we incur on mispredictions that would end up on the incorrect execution of the application. Therefore, these processors require special hardware in order to identify these mispredictions and recover the execution.

We describe the memory disambiguation pipeline of the Alpha 21264 as an example of this policy.

6.4.2.1 Case Study: Alpha 21264. This memory pipeline implements the following components in order to perform the disambiguation:

- **Load/Store Queue:** this queue holds memory operations until they compute their source operands and can be issued for execution.
- **Load Queue:** This queue stores the physical addresses of the loads in program order. A load allocates an entry on this queue at the renaming stage and reclaims it when it retires. This queue implements 32 CAM entries.
- **Store Queue:** This queue stores the physical addresses of the store instructions and its data in program order. A store allocates an entry at the renaming stage, and it does not reclaim it until retirement. This structure also implements 32 CAM entries.
- **Wait Table:** This table implements 1024 entries of 1 bit indexed by virtual pc. Whenever we identify that a load depends on a store it has overtaken, a store-load order trap is generated, and this table is updated setting the entry representing the virtual address of the load to 1. The fetch unit reads this table in order to tag the loads in case they have produced a memory violation in the past. Then, these loads would not be speculatively issued anymore. However, the wait table is reset every 16,384 cycles because we would end up with a table full of ones otherwise.

Loads and stores wait on the load/store queue until their source operands become ready. In the case of a store, we wait for the source operands required to compute the address and the data conversely to what happened on previous case studies. In case a load has its wait bit set, the load would not leave the load/store queue until all previous stores have been already issued.

Then, the source operands to compute the address are read in the next cycle, and the memory address is computed one cycle after.

Loads that have computed its address keep it on the load queue. Moreover, they compare its address with the address of younger loads, and in case they match, a load-load memory violation trap is triggered. This trap makes the processor to resume execution starting from the load that triggered the trap. If there is no need for a trap, the load proceeds to access the cache.

On the memory disambiguation stage, stores also write their memory address on their store queue entry. Moreover, they check the load queue looking for younger loads that match with its address. If this happens, a store-load memory violation trap occurs, and execution resumes from the load. Moreover, the wait table is updated in order to tag future instances of this load and avoid this situation to happen again. Note that stores do not check for store-store memory violations because even though stores are issued out of order, they do not update the cache until retirement. Since retirement occurs in program order, store-store memory violations never occur.

6.5   SPECULATIVE WAKEUP OF LOAD CONSUMERS

The latency of load operations is variable and mainly depends on whether the load hits on the TLB and data cache. There are also other factors that may affect the final cycle when the data will be available like, for instance, bank conflicts on the data cache, read port conflicts with other memory operations in the MOB, etc. However, the most common scenario is that loads provide data with the latency of a hit as shown in Figure 6.12. In this figure, we can see two possible scenarios where a consumer instruction reads a value produced by a load. As it can be seen, the load computes whether it hits on cache three cycles after select. Thus, if we implement a conservative wakeup where consumers are only woken up if it is guaranteed that the load will hit, we would obtain a two-cycle bubble between the producer and the consumer as shown in scenario 1. However, if we speculatively trigger the wakeup signal assuming that the load will hit as in scenario 2, we will be able to implement back-to-back execution for load operations as well.

However, in the infrequent case where a load misses in cache or its execution is delayed for some other reason, the consumers will have to be cancelled and reissued again.

As soon as an instruction leaves the issue queue, it is not guaranteed that this instruction may have an empty slot on this queue in order to go back. The issue queue may be full of instructions that indeed may be dependent on the instruction being reissued. Therefore, making instructions to wait for free slots on the issue queue in order to be reissued may end up in deadlocks.

There are several solutions to avoid this deadlock with different advantages and disadvantages. One solution is to flush all instructions in the pipeline younger than the one to be reissued and resume execution from there. This solution is similar to the aforementioned mechanism implemented on the Alpha21264 to recover from memory disambiguation misspeculations. The main drawback of this scheme is the significant performance drop we may have if this situation occurs very often. This is one of the reasons why the Alpha21264 implements the wait table.

Another solution that reduces the performance penalty of instruction reissue is to postpone the reclamation of the issue entry allocated by an instruction until we are sure this instruction would not have to be reissued. In this case, every issue queue entry has a bit (issued bit) that says whether

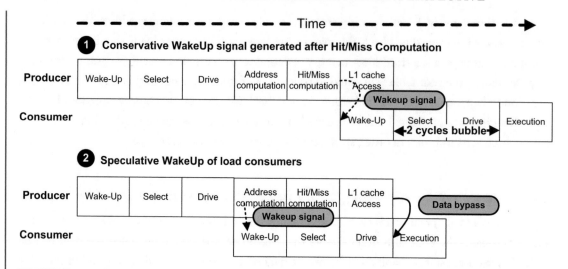

FIGURE 6.12: Load pipeline and consumer pipeline assuming (1) conservative wakeup and (2) speculative wakeup.

this instruction has been already issued. Then, issued instructions are not considered by the selection logic. However, whenever instruction reissue is needed, the issued bit is reset for the affected entries so that these instructions are considered by the selection logic again for execution. This mechanism reduces the penalty of reissuing instructions compared to the previous one, but increases the pressure on the issue queue. Note that all issue queue entries allocated by already issued instructions cannot be used to look for further independent instructions.

Unfortunately, the number of issue queue entries is usually small so that solutions like the previous one may degrade the performance due to the utilization of entries by issued instructions. The netburst architecture implements small issue queues in order to fit them into its tight cycle time. Moreover, since it implements a deep pipeline, the number of cycles instructions should stay on the issue queue since they are issued until we are sure they would not be reissued is very large. Therefore, netburst architectures like P4 implement a replay queue. In this case, instructions leave the issue queue (or scheduler) as soon as they are steered for execution, but they are queued in an additional fifo structure called the replay queue. Then, instructions reside in this queue until it is guaranteed they would not have to reissue again. However, if instruction reissue is needed, the scheduler gives priority to the replay queue in order to reissue its instructions in the order they were allocated there.

CHAPTER 7

Execute

It is in the execute stage that the program results are calculated. In this stage, an instruction's input operands (also known as source operands) are send to the processor's computational units along with the operation encoded in the instruction. The processor operates on the sources of the instruction and produces the result of the computation.

There are several types of operations that the processor can perform in the execution stage. The most common are the arithmetic operations (addition, multiplication, etc.). Memory instructions operate on data either by loading them from memory to registers or by storing them from registers to memory. Control-flow instructions change the value of the Program Counter (PC) register. More infrequently, specialized instructions can change the machine state by operating on control registers (special registers that define how the processor behaves).

Naturally, the different types of operations have different complexity and, as a consequence, different latency. For this reason, in contemporary microprocessors, the execute stage is not a single pipeline stage, but several. What is more, there are usually several different paths in the processor pipeline that an instruction can follow when it reaches the execute stage. The most obvious ones are the integer path, the memory path, and the floating-point path, with varying latencies. All these paths are consolidated at the write-back stage, when the results of the operation are produced and are written to the machine registers.

Most general-purpose out-of-order processors share the execution unit organization shown in Figure 7.1. The gray-shaded area in the figure shows the functional units (FUs) of the processor. The functional units correspond to the actual computation resources of the processor. In the figure, we can see four different types of units. The floating-point units (FPUs) perform arithmetic operations on floating-point values, as the name implies. The arithmetic and logical units (ALUs) are units that perform integer arithmetic operations and Boolean logic operations. The address generation units (AGUs) calculate the memory addresses for load and store instructions. Finally, the branch unit calculates the resulting PC value of control-flow instructions.

The register file in Figure 7.1 corresponds to all the architectural and in-flight register storage elements of the processor relevant to instruction execution. See previous chapters for possible organizations of this state (i.e., merged register file, architectural register file with values in the

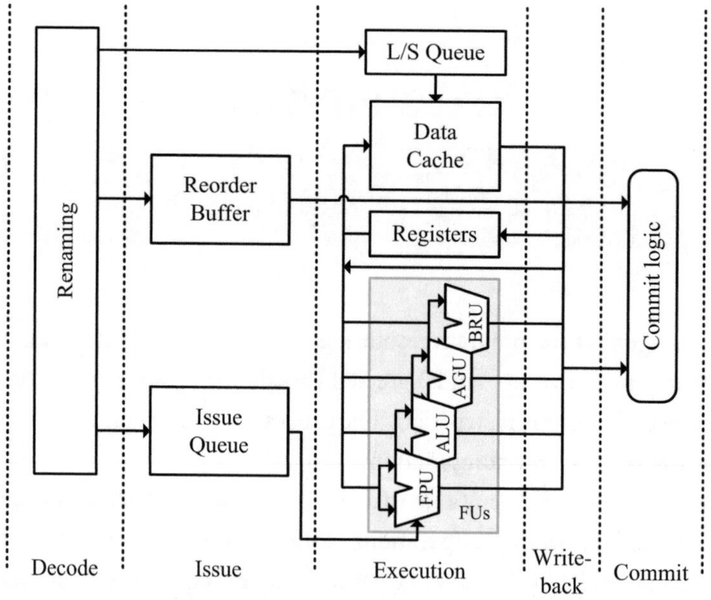

FIGURE 7.1: Processor execution units.

reorder buffer, etc.). Throughout this chapter, for simplicity, we will assume a merged register file organization that holds both architectural and in-flight values.

The data cache is another important part in the execution unit of the processor (see Chapter 2). The cache is used to provide fast access to frequently used data in memory and is an integral part of the load/store execution pipeline. The execution of load and store instructions implies the use of an address translation unit as well (not shown in Figure 7.1). The address translation unit is responsible for translating the virtual memory addresses encoded in the load/store instructions to the physical memory addresses that the Operating System has allocated for the program.

Another important aspect of the execution stage is the bypassing network. This is the network responsible for moving the sources and the results of the computation among the various functional units, the data cache and the register file. In modern microprocessors, some form of bypass is necessary if we want to provide back-to-back execution of dependent instructions. Because of its importance to performance and its complexity, the bypass network is one of the critical components of the execution stage.

Next, we describe the type of functional units typically found in contemporary processors, with special emphasis on functional units for multimedia support. Then, we describe several bypass network organizations: for a simple out-of-order machine, for a wide out-of-order machine and for an in-order machine. Finally, we study the design of clustered organizations, which have been used in some microprocessors to reduce power, area and the impact of wire delays.

7.1 FUNCTIONAL UNITS

The functional units found on a modern microprocessor can be classified based on the kind of operation they perform and on the type of data they operate on. In this section, we identify the different types of functional units, and we provide a description on the operations that each unit can perform.

7.1.1 The Integer Arithmetic and Logical Unit

This unit operates on two integer values coming from the general-purpose register file or the memory and produces an integer result. An integer arithmetic and logical unit (ALU) performs arithmetic operations such as integer addition and subtraction. Depending on the ALU, it can also perform integer multiplication and integer division. The ALU also performs bit-wise logical AND, OR, NOT and XOR operations. An ALU can also provide bit-wise NAND, NOR and XNOR operations, and logical operations where the second operand is inverted, such as ANDN, ORN and XORN. Finally, the ALU performs data transformation operations such as left or right shifting and rotation, and transposition (such as byte-swap) of the bits of one of its two operands.

Some instruction sets, such as the Intel x86 and the IBM POWER, implement condition codes or flags. In the x86 ISA, for example, there are six arithmetic flags: sign, parity, adjust, zero, overflow and carry. These flags are generated as a result of an arithmetic or logical operation, i.e., any operation performed in the ALU. For this reason, typically, the ALU of an x86-compatible processor has facilities to calculate the arithmetic flags along with the result of each computation.

7.1.2 Integer Multiplication and Division

Integer multiplication and division, although it operates on integer values, is not supported by the ALU. Due to the high complexity and area cost of the circuits needed for these operations, they are built as separate execution units in the processor (we call them here the IMUL and the IDIV units).

Moreover, in order to save area and power, many processors do not implement these units. Instead they use the floating-point unit (FPU) to perform integer multiplication and division. The Intel Atom processor is one such processor [15,23]. To perform an integer multiplication this way, first, the integer sources are converted to floating-point values, then the two numbers are multiplied and finally the result is back-converted to integer to produce the final instruction result. This process implies higher latency for the multiplication compared to having an IMUL unit, but depending on the applications, it may be worth the power and area savings (typical applications have very few integer multiplication and division instructions).

7.1.3 The Address Generation Unit

Memory instructions normally express the memory address that they want to operate on as a function of several source operands. It is the purpose of the address generation unit (AGU) to generate a direct pointer to the address space of the program from the operands of a memory instruction. The operation of the AGU heavily depends on the memory model supported by the machine. There are two commonly used memory models in microprocessors today: flat and segmented memory.

In the flat memory model, memory appears to the program as a single continuous address space. We call this space the *linear address space*, and flat memory addresses are called *linear addresses*.

In the segmented memory model, memory appears to the program as a collection of independent address spaces, the segments. A segment defines a single continuous address space starting at the *segment base address*. In this model, the program issues *logical addresses* to access memory. A logical address consists of a segment identifier and an offset inside the segment. The memory system of microprocessors internally uses only linear addresses, so the program logical addresses have to be translated to linear addresses by the processor's AGU. This is done by adding the segment base address and the segment offset part of the logical address together.

In both memory models, the linear part of the address (i.e., the linear address in the flat model and the offset part of the logical address in the segmented model) is called the *effective address* and is expressed as a function of one or more instruction operands. The *addressing mode* of an instruction defines how the source operands are combined to produce the effective address.

One of the ISAs with the most complex AGU requirements is the x86 ISA. The x86 follows a segmented memory model and has six different addressing modes. In x86, an effective address computation consists of the following components:

- *Displacement*: an immediate value encoded in the instruction bits.
- *Base*: the value of a general-purpose register.
- *Index*: the value of a general-purpose register.
- *Scale*: The constant 1, 2, 4 or 8 encoded in the instruction bits. The scale value is multiplied by the index value.

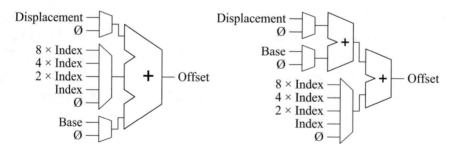

FIGURE 7.2: An x86 AGU offset calculation. The left circuit uses a three-input adder, while the right one uses a chain of two adders.

Wake-up	Select	Data read	Drive	Calculate offset / read segment info	Calculate linear address / check segment limit

FIGURE 7.3: Address calculation pipeline for x86.

The six x86 addressing modes allow for all the combinations of the following expression:

$$Offset = Base + (Index \times Scale) + Displacement$$

Figure 7.2 shows two different implementations of the offset calculation circuit for an x86 AGU. Calculating the offset is only part of the address calculation of the AGU. As mentioned earlier, the AGU must produce the linear address of the memory access by adding to segment base to the calculated offset. Moreover, the AGU must perform limit checks on the offset; that is, it must check that the offset is inside the boundaries of the segment.

As it can be seen from the above discussion, calculating the linear address for an x86 processor is a complex operation. Modern microprocessors that operate at high frequencies cannot perform such a complex operation in a single cycle. Figure 7.3 shows the execution pipeline stages for a possible implementation of such an AGU. It is a two-cycle process. In the first cycle, the offset is calculated, and the segment base and limit information is read from the segment register file.

In the second cycle, three actions take place. First, the base address is added to the calculated offset to form the final linear address. Second, the offset is compared to the segment limit to check if it points inside the segment. Third, the access permissions for the segment are checked (i.e., if a load operation has read access for the memory area defined by the segment).

Another possible implementation is to split the address calculation into multiple μops (e.g., one for doing base plus scaled index and another for adding segment base and doing limit checks). In this case, the AGU becomes simpler because it can be implemented with a simple adder, but memory operations generate multiple μops which may have some performance impact: we sacrifice some issue bandwith, and we cannot use the simple decoders (see Chapter 4) for some load operations.

7.1.4 The Branch Unit

The branch unit is responsible for executing the control-flow instructions (branches, jumps and function calls/returns) and for producing the correct next instruction address (we will call this the Program Counter or PC for short here).

Control-flow instructions can be conditional or unconditional. Conditional control-flow instructions (e.g., branches) change the flow of the program based on the result of a test (e.g., if two

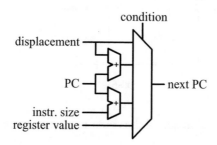

FIGURE 7.4: Next PC calculation in the branch unit.

register values are equal or if a condition code is set). If the test fails, the next instruction to be executed is the next instruction in memory. Unconditional control-flow instructions (e.g., jumps) always disrupt the program flow.

The destination PC of control-flow instructions can be defined in the following ways:

- **Direct absolute:** the instruction defines the next PC value explicitly.
- **Direct PC-relative:** the instruction defines the next PC value as an offset from the current PC (i.e., the PC of the control-flow instruction).
- **Indirect:** the instruction defines an integer register which contains the next PC value.

A branch unit thus has to be able to calculate the next PC for all the above cases. Figure 7.4 shows the next PC calculation part of the branch unit.

7.1.5   The Floating-Point Unit

This unit operates on two floating-point values coming from the floating-point register file or the memory and produces a floating-point result. A floating-point unit (FPU) performs arithmetic operations such as addition, subtraction and multiplication. Depending on the implementation, it can also perform division, square root and other complex operations (trigonometric functions, exponentials, etc.).

Normally, floating-point and general-purpose state is kept in separate register files. Depending on the architecture, there may be instructions that convert the floating-point values to integers and vice versa (and transfer the converted values from one register file to the other). Conversion operations are also implemented in the floating-point unit.

IEEE 754-1985 specifies four formats for representing floating-point values: (a) single-precision format which encodes values in 32 bits, (b) double-precision format which encodes values in 64 bits, (c) single-extended precision which encodes values in 43 bits or more and (d) double-extended precision format which encodes values in 79 bits or more.

In reality, most processors implement (a) and (b) in hardware. The Intel® x86 processors implement (d) as well using 80 bits for the encoding. The x86 extensions for floating-point numbers are also known as the x87 instruction set because they first appeared in the Intel® 8087 math co-processor.

The FPU is a very complex unit, and it is generally several times bigger than the integer units. For example, on the Pentium Pro, the FPU area is the same as the total area of 2 AGUs, 1 ALU, 1 IMUL and 1 IDIV unit [14].

7.1.6 The SIMD Unit

SIMD stands for single instruction multiple data, and as the name denotes, SIMD instructions are instructions that perform the same operation on a group of elements in parallel. SIMD instructions operate on SIMD registers. Normally, general-purpose and SIMD states are kept in separate register files. An example of the SIMD execution model is shown in Figure 7.5. Here we see the semantics of a SIMD instruction "z = x + y," that sums two 4-element vectors (x, y) to a third 4-element vector (z).

The first SIMD machines were the vector machines of the 70's (ILLIAC IV, CDC STAR-100 and Cray-1). These machines were designed to work on very large vectors (1-dimensional arrays of elements) with a single instruction. In these vector machines, it was typical to operate on vectors of hundreds of elements directly from memory. A vector addition such as the "z = x + y" example from above would sequence the elements of the two vectors to be added (x and y) from memory a few at a time to the processor's execution units. Computation and memory accesses are overlapped in this model, and very high performance can be achieved on data parallel computations, such as scientific computing.

In today's machines, SIMD has a very different form. Modern SIMD machines are designed to work on short vectors, and modern SIMD instruction sets usually set an architectural vector

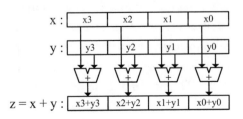

FIGURE 7.5: Parallel addition in the SIMD model.

length of 4–16 elements.[1] One of the long vector operations of the 70's machines would require multiple SIMD operations on today's machines. This difference in philosophy stems from the fact that the original reason for modern SIMD instruction sets was to speed up multimedia applications and games rather than scientific computing. Because of this implementation (and philosophical) difference, the term "vector" usually refers to vectors of many elements, while "SIMD" usually refers to vectors of a few elements.

The most popular SIMD instruction sets today are the x86 SSE [19,20] and the POWER AltiVec [17] extensions. Here we will focus on the x86 as a proxy for all modern SIMD extensions. The Pentium processor with MMX™ technology was the first x86 machine to introduce SIMD extensions to x86, focused on integer operations. The AMD K6-2 with 3DNow! [3], and later the Pentium III processor, followed with the streaming SIMD extensions (SSE) to x86, focused on floating-point operations. The first complete set of SIMD extensions for the x86—and the most used by compilers today—was SSE2 introduced in the Intel Pentium 4 and AMD Opteron processors. Newer versions of SSE (SSE3, SSSE3 and SSE4.x) did not introduce any new data types to SSE2, only new instructions.

The SSE x86 extensions define 16 new SIMD registers to the x86 ISA of 128 bits each. Similarly, the AltiVec extensions define 32 new 128-bit SIMD registers to the POWER ISA. Each SIMD register can represent vectors of different types of elements as follows:

- A vector of 16 byte-sized (8b) elements.
- A vector of 8 word (16b) integer elements.
- A vector of 4 doubleword (32b) integer elements.
- A vector of 2 quadword (64b) integer elements.
- A vector of 4 single-precision floating-point (32b) elements.
- A vector of 2 double-precision floating-point (64b) elements.

The implementation of a SIMD unit must support all the vector types and operations defined by the ISA. In SSE, apart from the arithmetic operations, the ISA defines bit-wise logical operations on the entire vector registers and element permutation operations (shuffles).

Usually the SIMD unit is composed of individual sub-units: a floating-point unit, an integer/logical unit and a shuffle unit. Moreover, each unit is further divided into *lanes*. A lane is the minimum building block of a vector unit, i.e., a circuit that can operate on two vector elements and produce a one-element result. Remember that SIMD operations are parallel, so they are independent of each other. Thus, a SIMD unit can be built just by putting next to each other multiple

---

[1]Vector length is typically measured in number of single- or double-precision floating-point elements.

copies of the same lane. Moreover, the hardware for an integer SIMD lane is almost identical to the ALU hardware, while the hardware for a floating-point SIMD lane is very similar to the FPU hardware (except for the support for 80-bit operations in x87), which helps when we want to reutilize components.

The width of the "scalar" execution units such as the ALUs and the FPUs is the same as the width of the architectural register (64 bits for the ALU and 80 bits for the FPU in x86), but we do not have this restriction in the SIMD unit. Given our definition of a lane, for a SIMD instruction set where the register size is N-lanes wide, we could implement the SIMD unit by utilizing anywhere from 1 to N lanes. In SSE, since the largest vector element is 64 bits, lanes are also 64 bits, so an SSE unit can be built with either 1 or two lanes side-by-side.

The two alternatives for the SIMD unit (1 lane vs. two lanes) are shown in Figure 7.6. In this figure, we show the execution of two SSE instructions "a = x op y" and "b = z op w" that operate on 32-bit elements. In this case, each lane can perform two operations in parallel. On the top part of the figure, we see the operands of two instructions. In the middle part of the figure, we see how these two instructions are executed when we have two lanes. Basically, we have enough execution bandwidth to operate on an entire 128-bit SSE vector per cycle.

An alternative design with lower performance but with half the hardware cost is shown in the bottom part of the figure. Here we only have 1 lane, so we split one SSE operation into "low"

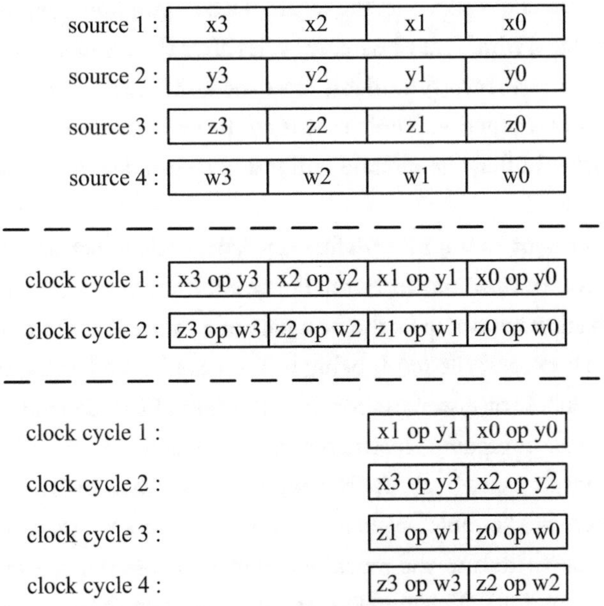

FIGURE 7.6: One-lane (bottom) vs. two-lane (middle) SSE unit.

and "high" parts. The SIMD unit will operate on the low part in one cycle and on the high part on the other. The issue bandwidth in this case is reduced by two (but latency per instruction has only increased by one cycle).

As mentioned earlier, one floating-point SSE lane is very similar to an FPU. It makes sense for an x86 processor to share the same hardware between the FPU and the SIMD unit by designing a hybrid unit that can do one 80-bit floating-point operation, or one 64-bit floating-point operation, or two 32-bit floating-point operations. This unit can operate as an x87 FPU in "scalar" mode (only one of 80/64/32-bit operations) or as an SSE floating-point lane (one 64-bit or two 32-bit operations).

The number of lanes does not have to be the same for all sub-units of the SIMD unit. For example, a design can implement a single floating-point lane, but two integer lanes. What is more, if a design implements multiple lanes of the same type, they do not have to be symmetrical. It is not uncommon, for example, a two-lane floating-point SIMD unit, to have only one lane for divisions or other complex operations, trading off some performance for significant area and power reduction.

One example of a non-uniform SIMD unit is the Intel Atom processor [15]. In Atom, the floating-point SIMD unit is single lane and shared with the FPU. The SIMD shuffle unit is also single lane, but the integer SIMD unit has two lanes.

7.2   RESULT BYPASSING

When executing instructions in a pipeline, the result of a computation does not update the machine state until the commit stage, which may be many cycles after the result was generated. The result of the computation becomes *speculatively* available after the write-back stage, though. The write-back stage is when the result of a functional unit is sent to the architectural register file, to the merged register file, to the reorder buffer, the rename buffer and so on, depending on the machine design (in-order, out-of-order, etc.).

In a pipelined processor, to improve performance, dependent instructions can execute *speculatively* by reading their source operands from the noncommitted machine state. In today's pipelined processors, the write-back stage occupies the best part of a cycle. The same occurs for the read operand operation. In the best case, the result being written can be read at the same cycle.

Figure 7.7 shows a best case scenario for the execution of two dependent instructions. In this figure, we assume a very short pipeline: instruction issue (wakeup and select), operand read, execute (ALU operation) and result write-back. As it can be seen, the two instructions cannot execute in consecutive cycles—there is a "bubble" cycle in between—because we have to wait for the result of the first instruction to be written to the machine registers before the second instruction can read it (it's an input to its operation). This pipeline can be representative of a low-frequency in-order machine or out-of-order machine.

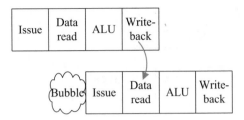

FIGURE 7.7: Dependent instruction execution in a simple pipelined processor.

A more typical pipeline for a high-frequency out-of-order execution core is shown in Figure 7.8. Here we assume a pipeline with a merged register file. Similar to the discussion in Chapter 6, we assume that the source operand values require a whole cycle to reach the ALUs after they are read from the register file. The result also requires a whole cycle to travel from the output of the ALUs to the register file. In this case, the minimum number of lapsed cycles between two dependent instructions is four.

In the two examples above, it can be seen that we are losing performance if we have to wait for the result of an operation to be written to the register file before the dependent instructions can use it. Of course, the compiler can alleviate some of this by reordering program instructions in such a way that dependent instructions do not appear back-to-back. This is not always possible, though. It is also evident that an in-order pipeline will suffer more since an out-of-order processor can dynamically schedule independent instructions to fill the bubbles.

The hope is that for a pipeline such as the one shown in Figure 7.7, the compiler and the out-of-order engine can fill most of the bubbles with useful instructions, but fully utilizing the pipeline of Figure 7.8 is an almost impossible task: for a single issue processor, the program must have a sustained instruction-level parallelism (ILP) level of four, i.e., there must be four (almost) independent streams of instructions in any given point in time.

The immediate observation by looking at Figures 7.7 and 7.8 is that we do not really need to wait for the result to be written back to use it; the correct value is available at the end of the ALU stage of the producer. Also, the consumer will not really use the data until the beginning of its ALU

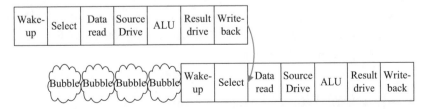

FIGURE 7.8: Latency for dependent instruction execution in a deeply pipelined processor.

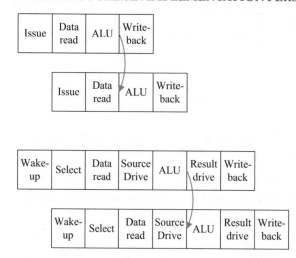

FIGURE 7.9: Dependent instruction execution with value bypassing in a simple pipelined processor (top) and in deeply pipelined processor (bottom).

stage, so we could build a pipeline that *forwards* or *bypasses* values directly from the execution (ALU) stage of an instruction to the execution stage of the next, as shown in Figure 7.9.

In this case, two dependent instructions can execute back-to-back with no bubbles. This significantly improves performance, but it requires a new data path in the processor dominated by wires and multiplexors (that is why it is typically called the *bypass network*). Depending on the pipeline, the bypass network can be from relatively simple to very complex. In all cases, though, the bypass is one of the most critical aspects of the execution engine design because it affects the area, power, critical path and physical layout of the execution stage.

Implementing a bypass network, like most other aspects of processor design, is a tradeoff. Having bypasses improves the executed instructions per cycle metric (IPC), but it may affect the cycle time and/or power of the microprocessor. Most processors today implement some form of bypass. The notable exception is the IBM POWER4 [42] and IBM POWER5 [38] processors, where the designers opted to not implement a bypass network in order to keep complexity low (and frequency high). In these machines, executing two dependent integer instructions requires a one-cycle bubble, while executing two dependent floating-point instructions requires six cycles of bubbles. Nondependent instructions of course can execute during these bubble cycles since both of these processors employ out-of-order execution.

7.2.1   Bypass in a Small Out-of-Order Machine

Figure 7.10 shows a simple execution engine of two functional units. In this case, we assume a low-frequency machine, with a shallow pipeline (Figure 7.9 top). Our design also assumes a merged

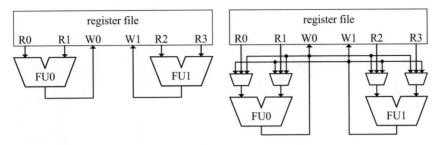

FIGURE 7.10: Simple execution engine with two functional units, without (left) and with (right) value bypassing.

register file, although the bypass will work the same in the case of an ROB-based write-back. Comparing the no-bypass design (Figure 7.10 left) to the design with a bypass network (Figure 7.10 right), we can immediately see how the complexity of the execution engine increases.

In the no-bypass case, each input of a functional unit is connected directly to a read port of the register file to read the source value. Similarly, the result of the functional unit is connected directly to a write port of the register file. If we want to implement value bypassing, the source value of a functional unit can come from three different places in the machine in this design: the register file (i.e., no bypass), the functional unit itself and other functional unit. Thus, we need a 3:1 multiplexor at the input of each functional unit. Also, the results of the functional units, instead of connecting directly to the register file, now form a bus that spans the width of the execution engine (called the *result bus*) and connect to all the functional unit input multiplexors.

7.2.2 Multilevel Bypass for Wide Out-of-Order Machines

It is obvious from the above discussion that a bypass network increases the complexity of an execution engine. What is also true is that the more complex the execution engine (more functional units, deeper pipeline, etc.), the more complex the bypass network. In Figure 7.11, we show a deeply pipelined execution engine with two functional units. In this case, we assume a machine with the same pipeline as in Figure 7.9 (bottom). We also assume a merged register file, although the bypass will work the same in the case of an ROB-based write-back.

At the left of Figure 7.11, we can see the execution engine without value bypassing, and at the right, we can see a possible implementation of a bypass network for this machine. In the no-bypass case, each input of a functional unit is connected directly to a read port of the register file (through a latch) to read the source value. Similarly, the result of the functional unit is connected directly to a write port of the register file (again, through a latch). At the right of Figure 7.11, we can see the execution engine with value bypassing. In the following paragraphs, we explain the details of this design.

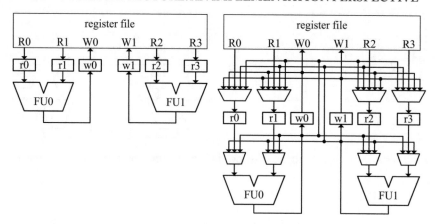

FIGURE 7.11: Deeply pipelined execution engine with two functional units, without (left) and with (right) value bypassing.

This case is different from the one of the previous section; depending on the distance in time (in clock cycles) the producer and the consumer are scheduled, the data can be forwarded from different stages of the producer to different stages of the consumer. Figure 7.12 shows the possible forwarding paths for this pipeline from an instruction I1 to instruction I2–I4 scheduled in different cycles. Here we assume that write-back and data read can overlap to simplify the discussion. For this design, we have also assumed that we forward from the producer to the consumer as soon as possible. For example, an alternative for the "ALU" to "source drive" bypass would be a path from "result drive" to "ALU."

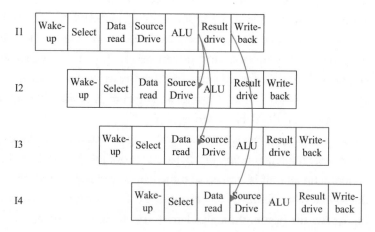

FIGURE 7.12: Value bypassing paths for a deeply pipelined execution engine.

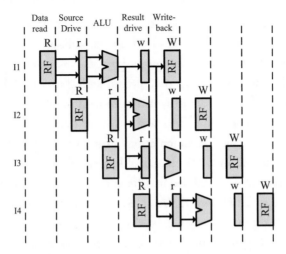

FIGURE 7.13: Value bypassing data transfers in a deeply pipelined execution engine.

In Figure 7.11, the "r" latches are placed at the end of the "source drive" and at the beginning of the "ALU" pipe stage. Also, the "w" latches are put at the end of the "result drive" and at the beginning of the "write-back" stage. Not shown in Figure 7.11 is the output latch of the FU that latches data at the end of the "ALU" stage to the beginning of the "result drive" stage. Given this design, in Figure 7.13, we can see the paths shown in Figure 7.12 between instruction I1–I4 and which hardware blocks are involved. In the figure, we only show the pipeline latches, the register file and the ALU. The only paths shown are the data flow of I1 and the bypasses from I1 to I2–I4. By looking at this figure, it is easy to derive the design of Figure 7.11 (right).

7.2.3   Bypass for In-Order Machines

Although it is generally believed that in-order machines are simpler than out-of-order ones, this is not necessarily true for the result bypassing part of the processor. In-order machines must delay the write-back stage of execution until the slowest FU operation in the pipeline has finished (e.g., a multiplication, a memory access, etc.); otherwise, we could produce architectural state out-of-order. The process of delaying the result write-back is called *staging* of the result (the latches involved are called the staging latches). Some form of staging takes places in out-of-order machines also, for example, the "result drive" stage of Figure 7.12, but in-order machines—especially high-frequency ones—may employ deeper staging (Intel Atom has two to five cycles of staging [15], and ARM Cortex A8 has between one and three staging cycles [4], depending on the operation). If we do not want to introduce bubbles, we must forward results from all staging latches in addition to the outputs of the FUs.

As an example, in Figure 7.14, we show the pipeline of the Atom processor [15]. The Atom pipeline is optimized for the x86 ISA that has load-op instructions, that is, instructions where one of the operands can come from memory. For this purpose, the execution pipeline of Atom has, first, the AGU stage, then the data access stage (two cycles) and, third, the ALU stage. Write-back is staged for two cycles after ALU in order to synchronize with the floating-point execution pipeline (not shown) to check for exceptions and faults and to handle multithreading.

Although we have no detailed documentation of the Atom bypass design, in Figure 7.14, we show a reasonable implementation for all the possible forwarding paths for this pipeline from an instruction I1 to instruction I2–I7 scheduled in different cycles. As shown in the figure, AGU, load or ALU results can be bypassed to both the AGU and ALU. This means that an AGU input can come from seven different pipeline stages. Comparing this single-issue example to the 2-way super-scalar, out-of-order forwarding paths in Figure 7.12, we can immediately appreciate the similarity in complexity of the in-order bypass (Atom is 2-way superscalar as well, but we do not explore this dimension here for simplicity).

Given this design, in Figure 7.15, we can see the paths shown in Figure 7.14 between instruction I1–I7 and which hardware blocks are involved. The only paths shown are the data flow of I1 and the bypasses from I1 to I2–I7. Here we have made some assumptions in our design. First, we cannot have too many inputs directly into the ALU/AGU sources (the input multiplexor into the ALU can affect cycle time if it becomes too big). That is why we try to bypass into the previous pipeline stage (in "data read" for AGU, for example). Second, we assume that once bypassed, data

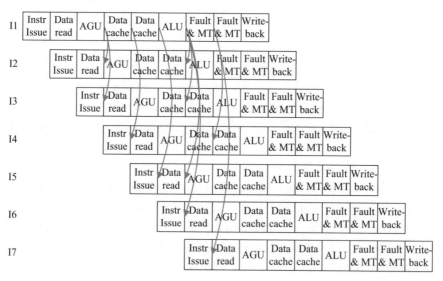

FIGURE 7.14: Possible value bypassing paths for the Intel Atom in-order execution engine.

travel horizontally through the correct channels in the pipeline. This means that a result forwarded at the AGU stage can be consumed in the ALU stage of the instruction (e.g., the forwarding of the I1 ALU or AGU result to the I5 ALU operation).

Another difference between out-of-order and in-order result forwarding is the storage elements required. For example, out-of-order machines use a merged register file (or ROB or rename buffers) to hold results not yet committed to architectural state (in-flight results). The bypass network is thus either (a) direct FU result to FU source or (b) RF/ROB to FU source. In-order machines use the staging latches for holding in-flight results. Figure 7.16 shows a conceptual block diagram for the bypass of Figure 7.15. In this figure, A and B are the FU sources, and C is the FU result.

A straightforward implementation of the staging latches would implement one latch per result per staging cycle. This implementation would then require a different data bus for each staging latch that we want to forward results from. This creates a significant wiring problem for the bypass.

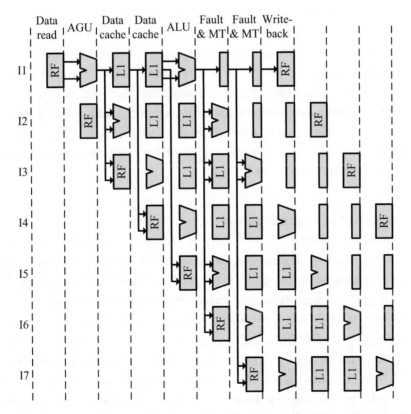

FIGURE 7.15: Possible value bypassing data transfers for the Intel Atom in-order execution engine.

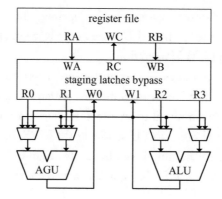

FIGURE 7.16: In-order execution engine with AGU and ALU, with AGU-to- ALU and ALU-to-AGU/ALU value bypassing.

Another issue with this implementation of staging is that data gets copied from one latch to the next every cycle. This consumes a lot of unnecessary power.

The preferred way to do the staging is by implementing a register-file structure with as many entries as the depth times the width of the execution pipeline (let us call this the SRF for the rest of our discussion). At the instruction issue time, an SRF entry is allocated to hold the instruction result. The result is written into the SRF as soon as it is produced and waits there until the write-back stage when it is copied into the architectural state. This design is conceptually the same as having an ROB structure to hold in-flight results (although its complexity is much reduced). Bypassing in this scenario becomes very similar to the out-of-order case then: results can come either directly from the FUs, or the architectural state or the SRF.

7.2.4 Organization of Functional Units

From the discussion in Sections 7.2.2 and 7.2.3, it is evident that for modern microprocessors that have high-frequency, wide execution engines (i.e., many functional units), it is not possible to bypass values from any functional unit to any other: the bypass network would grow too large (even the two FU network in Figure 7.11 is not simple) and would impact the cycle time of the machine.

Luckily, we do not have to interconnect all functional units. In most microprocessors, for example, the floating-point and SIMD units have their own bypass network that does not connect the integer part of the execution engine. The rationale behind this separation is that integer operations very rarely, if ever, have sources that are the result of a floating-point operation and vice versa. On the other hand, address calculations often utilize the results of integer instructions, so it is logical that the AGUs and the ALUs of the execution engine are connected in a single bypass network.

One unit that we have not discussed so far is the memory. In Chapter 2, we discussed the cache memory in more detail, but as far as the bypass is concerned, it is just another set of functional units: load operations generate results that are put in the result bus, while store operations read data off of the result bus. Since memory instructions are important in both integer and floating-point/SIMD operations, usually, the memory is connected to both the ALU/AGU and the floating-point/SIMD bypasses.

7.3 CLUSTERING

The latest generations of high-performance superscalar microprocessors have increased significantly the width, depth and amount of speculation in the pipeline. All of these trends have increased performance many-fold, but they come at a significant increase in hardware complexity as well. Other limiting factors for continuing in this direction—apart from complexity—are power and temperature, and the scalability of global/long wires such as those found in bypasses and multiported array structures.

A design philosophy that has proven effective is to partition the layout of critical hardware components—whenever feasible—so as to maintain most of the parallelism while improving the scalability. Examples of this technique are the array replication in caches, explained in Chapter 2, and the distribution of the issue logic (see Chapter 6). Clustered architectures extend this divide and conquer philosophy into all the execution core resources, such as the register file, the issue queue and the bypass network. By its nature, clustering can be applied to different levels, with varying granularity. Here we explain different clustered designs, from most "conservative" to the most aggressive.

7.3.1 Clustering the Bypass Network

This is the simplest form of clustering. This design may become necessary when the complexity of the bypass network increases so that it can affect the cycle time of the processor. In Figure 7.17, we can see an example of how this works. On the left side, we can see a non-clustered multilevel bypass design, typical of a deeply pipelined processor (this is the same as the right part of Figure 7.11).

By clustering the bypass, we do not allow values to be forwarded from FU0 to FU1 and vice versa, although we do have a local bypass to each functional unit. Unlike the no-bypass case (left of Figure 7.11), we can issue dependent instructions without bubbles if they go to the same functional unit. When dependent instructions are issued to different functional units, though, communication happens through the register file, which will incur bubbles.

As can be seen in Figure 7.17, this design has much less complexity. We have much less wires, and we have removed a whole pipeline stage from the execution core. This means lower power

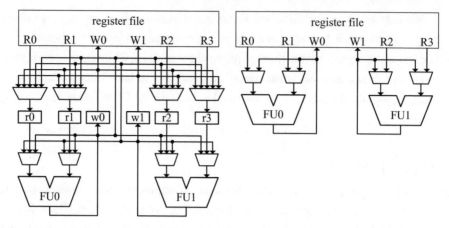

FIGURE 7.17: Simplifying a multilevel bypass using clustering (right). On the left, we can see the non-clustered design.

consumption and higher execution frequency, which may result to a net benefit, even though we introduce some bubbles in the pipeline.

7.3.2   Clustering with Replicated Register Files

The register file is a challenge for a wide machine if we want high frequency. As the number of read and write ports of the register file increases, the access latency increases. The designers of the Alpha 21264 processor were faced with this issue [25]. The solution they decided to adopt was to cluster the design. In the execution engine, there are a total of four integer units, divided into two clusters. Each cluster also includes a copy of the register file. The two copies are kept coherent by broadcasting the values of the functional units to both register files. Broadcasting requires one extra cycle for a value in cluster 1 to be available if it is produced in cluster 0 and vice versa.

The Alpha 21264 uses a unified issue queue for all four integer units (i.e., for both clusters). The issue queue uses two arbiters, one for each cluster to decide where instructions will be issued. Each instruction is statically assigned to an arbiter upon entering the issue queue, based on instruction fetch position [25]. The goal of the arbiter assignment algorithm is to balance the utilization of the two clusters.

Figure 7.18 shows a simplified block diagram of the Alpha 21264 execution engine. In the figure, we do not show the load and store buses, and their connections to the bypasses. This design compared to a nonclustered one has half the read ports in the register file and much simplified bypass, which allowed the Alpha 21264 to reach its very aggressive (for the time) execution frequency.

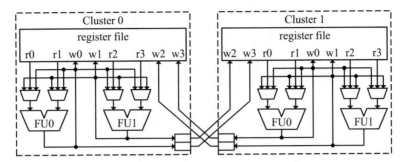

FIGURE 7.18: Simplified block diagram for the Alphsa 21264 execution engine, showing the two clusters.

The unified issue queue does not reduce wakeup complexity but allows delaying the cluster assignment, which is most effective to avoid inter-cluster communication penalties and to balance the load of the two clusters.

7.3.3   Clustering with Distributed Issue Queue and Register Files

More aggressive clustered architectures have been proposed that do not replicate the register file, but instead they distribute it. That is, instead of two copies of the entire register file, like in the Alpha 21264, the processor has one register file but divided into two parts, each half the size (both in number of ports and number of entries). Distributed register file clustered architectures distribute the entire execution datapath, the issue queue and the issue logic as well. Some of the earliest works on this type of clustering include the multicluster architecture [11] and the work of Canal *et al.* [6,7] and Zyuban [50] and Zyuban and Kogge [51].

With a distributed register file, only one copy of each physical register is kept in the system, and an instruction does not broadcast its results, but it writes to its local register file only. Also, an instruction can only read its operands from the local register file. The architecture provides explicit mechanisms to communicate values from one register file to the other. One such mechanism is by maintaining a small register file with registers that are replicated; another is by providing a special "copy" instruction that does an intercluster register-to-register move.

Another characteristic of this type of clustered architecture is that the issue queue is also distributed and is local to a cluster. Instructions are assigned a cluster at the rename stage through an *instruction steering* mechanism, and then they only compete for resources in that cluster. Intercluster dependences among instructions are handled at the rename stage.

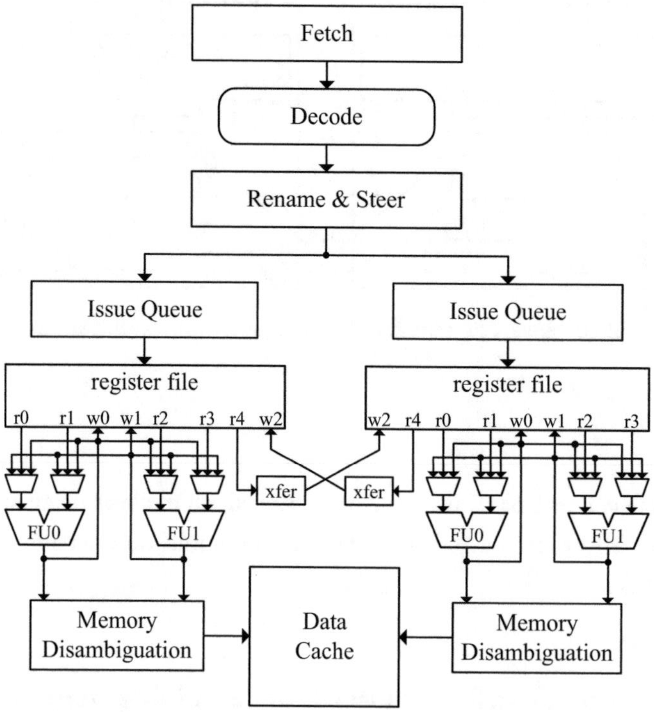

FIGURE 7.19: Simplified block diagram for a multicluster-type design, showing two clusters.

Figure 7.19 shows a high-level diagram of this type of clustering. This design reduces power consumption and complexity compared to the previous ones, but it puts significant pressure on the instruction steering mechanism. The best steering mechanisms try to reduce the expensive intercluster communications by steering dependent instructions to the same cluster, while at the same time, they try to balance the workload among the clusters in order to avoid unnecessary resource-induced stalls (one cluster saturates its resources, while another one is idle).

·   ·   ·   ·

CHAPTER 8

The Commit Stage

8.1 INTRODUCTION

Most current processors are based on an execution model based on sequence of instructions where one instruction is executed right after the previous one completes. Therefore, processors behave as if they would be executing instructions one after the other in the original sequential order [41]. However, neither in-order nor out-of-order processors really begin executing an instruction right after the previous one completes. Current processors are pipelined so that they always have several instructions in-flight at different phases of their execution. Thus, instructions may modify the processor state in an order different than the sequential order. For instance, if we execute a sequential program comprising instructions A and B where A is older than B, even in an in-order processor, it may happen that if A triggers an exception at the final stages of its execution, B may have already modified some register value if it had time to reach the write-back stage. In this case, the processor would not be able to provide the architectural state after A and before B to process the exception.

The most common solution for existing processors in order to emulate the sequential execution of instructions is to implement an additional stage called commit at the end of the pipeline. Instructions flow through this stage in the original program order. Then, any changes instructions do on previous pipeline stages are considered speculative and do not become part of the architectural state until they reach commit. At this point, we say that the instruction commits.

In the previous example, the exception triggered by A would be handled when A commits. Moreover, since instructions are committed in order, B would not have committed yet so that its changes would not be part of the architectural state. This way, the exception produced by A could be handled as if any instruction younger than A would have never been executed. We say that a processor supports precise exceptions if it provides the correct architectural state as it was before the execution of the instruction that triggers an exception.

A processor operates with two separate states: the architectural state and the speculative state. The architectural state is updated at commit as if the processor would execute instructions in sequential order. By contrast, the speculative state implies the architectural state plus the modifications performed by the instructions that are in-flight in the processor. This latter state is called speculative because it is not guaranteed that these modifications will become part of the architectural state. Note that conventional processors rely on speculative techniques like branch prediction or speculative memory disambiguation in order to keep executing instructions. Thus, if some of these speculations fail or an exception occurs, the speculative state becomes invalid, and it never turns into architectural state.

In contrast to RISC processors like Alpha or MIPS, some CISC processors like the latest x86 processors (Intel Pentium III, Intel Pentium 4, Intel Pentium M or Intel Core architecture) split the x86 instructions into simple micro-operations in order to facilitate the implementation of the out-of-order engine. In this case, an x86 instruction commits and updates the architectural state of the processor when all the micro-operations belonging to the x86 instruction have successfully completed their execution. The only exceptions to this rule are those x86 instructions that are split into a large number of micro-operations like memory copy instructions. In this case, the instruction does periodic partial commits at specific points of its execution since this is accepted by the x86 semantics as detailed in the Intel reference manual [19].

Finally, since the commit is the last stage on the execution of an instruction, this is the place where execution hardware resources allocated by the instruction like reorder buffer (ROB) entries, memory order buffer (MOB) entries or physical registers are reclaimed. Note that an instruction should only reclaim those resources that are not used anymore. Therefore, for those configurations where the instructions write their outcome in a physical register, the reclamation of this physical register should be done by the time we know for sure that the content of the register would not be needed anymore. Thus, before reclaiming a register, we need to be sure that all instructions that may require the value of this register in the future have already read it or they will be able to read it from a different place.

In the next sections of this chapter, we describe some alternatives in order to update the architectural state. Moreover, we also explain several mechanisms in order to recover the speculative state and resume execution when branch mispredictions or exceptions occur.

8.2 ARCHITECTURAL STATE MANAGEMENT

The architectural state comprises the memory state plus the value of every logical register.

As part of the architectural state, the memory cannot be modified until an instruction commits. The reason is that we cannot propagate memory updates to the rest of the system (devices, other cores, etc.) until we are sure that the updates are correct. Thus, all store operations do not update the memory state until they commit. In the meantime, they reside in an entry of the store buffer along with the memory address they modify, the size of the modification and the value to store in memory at commit. Some processors like the PA8000 or the MIPS R10000 do not store the data on the store buffer, but they read it from the register file using a dedicated port at commit time.

Therefore, all load operations should check whether there is any older store in the store buffer that updates the memory space they read. In case a load finds an older store matching with the memory addresses it is going to read, the load should either get the data from this store instead of from the cache or wait for the store to update the cache. Further details regarding the memory management can be found in Chapter 6.

There are several ways of keeping track of the latest architectural state for the logical registers. These methods are also dependent on the allocation scheme implemented on the processor. In this section, we will cover two methods: the architectural state management based on a reorder buffer (ROB) and retire register file (RRF) like in P6 [39] or Intel Core, and the management based on a merged register file that holds the speculative and architectural values. This latter method is used in processors like the Intel Pentium 4, Alpha 21264 or MIPS R10000 [27,33,48].

8.2.1 Architectural State Based on a Retire Register File

Processors like the P6 implement a reorder buffer (ROB) where instructions store the produced values until they retire and become part of the architectural state. Then, the values are copied into a register file with as many entries as logical registers available. This register file stores the architectural state for every logical register and is usually called retire register file (RRF).

Figure 8.1 shows this scheme. The ROB is a circular FIFO where instructions allocate a new entry at the allocation stage (a.k.a. renaming stage), and this entry is reclaimed as soon as the instruction commits or is squashed due to branch mispredictions or exceptions, among others. Then,

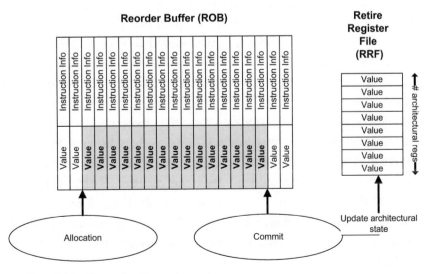

FIGURE 8.1: Reorder buffer and retire register file.

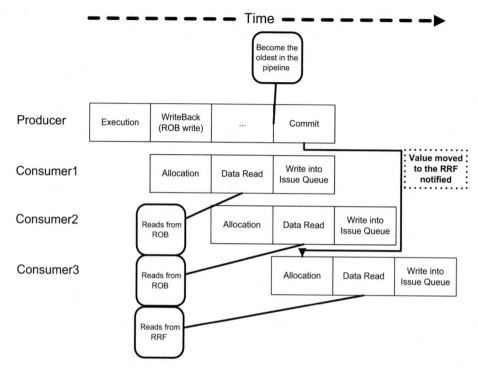

FIGURE 8.2: Notification of produced value transferred to the RRF.

every entry in the ROB includes two main sections: instruction information and the produced value. The instruction information implies data about the type of the instructions, the status of its execution and the identification of the architectural register they produce. On the other hand, the produced value field holds the value produced by the instruction. Therefore, the values produced by the in-flight instructions (shaded in the figure) represent the speculative state of the machine. As soon as an instruction commits, the architectural register modified by the instruction is updated in the RRF with the value stored in the ROB, and its ROB entry is reclaimed. Thus, the RRF always holds the authoritative copy of the architectural state of the processor.

Note, however, that the place where a value is stored on this scheme varies during its lifetime. First, the value is stored in the ROB so that consumer instructions will read it from there. However, as soon as the instruction that produced the value commits, its ROB entry is reclaimed, and the value is written into the RRF. When the value is moved to the RRF, all consumers that have not read it yet should know that it is not available on the ROB anymore but on the RRF instead. This fact could complicate the implementation of certain pipeline stages depending on the characteristics of the microarchitecture as described later.

Most of the ROB-based architectures read the operands before issuing the instructions, right after the renaming stage. This situation is shown in Figure 8.2. As it can be seen, all consumers read the produced value from the ROB until the producer instruction commits. At this cycle, the value still resides on the ROB entry since the entry has not been reused yet. Therefore, instructions reading their operands at this moment could still get it from the ROB. However, the allocation stage should be notified in order to update the renaming table and mark this value as available in the RRF. Note, though, that the renaming table is only affected by this change if any of its entries is still pointing to this value. As soon as the renaming table is updated, all consumer instructions renamed since then will know that the value is available in the RRF instead of in the ROB.

However, implementing an ROB-based architecture with RRF where consumers read their source operands after issue complicates this notification. In this case, the commit logic does have to notify the invalidation of the ROB entry not only to the allocation stage but also to all instructions residing in pipeline stages between allocation and issue, including instructions in the issue queue. Then, each of these instructions should check whether the notification affects to any of their sources and update their information accordingly in order to read their sources from the right place when they are issued.

8.2.2 Architectural State Based on a Merged Register File

Processors that implement a merged register file like the MIPS R10000, the Alpha 21264 or the Intel Pentium 4 use the same register file for the values belonging to the architectural state and the speculative values. Basically, a physical register is allocated by an instruction to store its produced value, and this register will hold this value until it is not needed anymore, even if the instruction commits. This characteristic offers three main advantages compared to ROB-based schemes.

Values do not change location when they commit. Thus, the renaming table and in-flight instructions that have not read their source operands do not need to be notified when a value becomes part of the architectural state. This makes this scheme suitable for processors where instructions read their source operands after issue.

A ROB-based implementation requires space for as many produced values as the number of in-flight instructions supported by the ROB. However, around 25% of the instructions in typical applications are usually store operations and branches that do not produce any value [16]. A register file exploits this feature by implementing a more power effective hardware structure where the number of registers to write the produced values is lower than the number of in-flight instructions supported by the architecture.

The ROB is a centralized structure that has to be accessed by all instructions that need to read their source operands. Therefore, this design is suitable if instructions read their source operands after the renaming stage where the instruction management is still centralized. However, an ROB would complicate decoupled architectures where instructions read their sources after issues like the Intel Pentium 4, MIPS R10000 and Alpha 21264. These processors steer the instructions to different execution clusters depending on the type of resources it needs for execution. In these cases, implementing centralized structures like the ROB where decentralized hardware like the execution clusters should access is not recommended. A register file-based architecture is more suitable for these latter scenarios where separate register files could be implemented on every execution cluster.

The first advantage is related to the fact that the speculative state and the architectural state share the same hardware structure. However, the other two advantages come from the fact that we use a physical register file to store the values instead of an ROB. Therefore, we could observe similar advantages using other schemes to keep track of the architectural and speculative states like future files [41].

By contrast, a merged register file, as any other hardware scheme that relies on a register file, complicates the renaming compared to an ROB-based architecture. In a nutshell, whereas ROB-based architectures take advantage of the ROB entry that is assigned sequentially to store the produced values, the register file-based architectures require an additional list on renaming that stores the register file identifiers that are available. A more thorough description of the implications on the renaming stage can be found in Chapter 5.

The resource reclamation is also more complicated on this scheme. Whereas the ROB entries are reclaimed sequentially as soon as the instruction commits, a merged register file cannot reclaim any physical register until it is guaranteed that the value it holds would not be needed anymore. Processors that implement a merged register file use a conservative approach to reclaim physical registers. In general, a processor reclaims a physical register allocated by instruction A when another instruction B younger than A that writes the same logical register as A commits.

In the next section, we describe the way the speculative state could be recovered, and execution resumed in the event of a branch misprediction or exceptions.

8.3 RECOVERY OF THE SPECULATIVE STATE

One of the reasons why in-flight instructions are not finally committed is because they were fetched in a wrong path due to branch mispredictions or a younger instruction raised an exception.

In any of the two cases, the speculative state should be recovered to undo the modifications produced by the renaming of the instructions that would not be committed. The next two sections describe typical recovery mechanisms used in case of branch mispredictions and exceptions. These schemes depend on whether the processor design is based on an ROB with a separate RRF or a merged register file.

8.3.1 Recovery from a Branch Misprediction

In the event of a branch misprediction, the speculative state of the machine is incorrect because it has been fetching, renaming and executing instructions from a wrong path. Therefore, when we

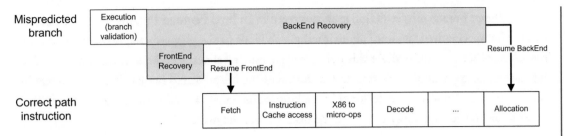

FIGURE 8.3: Generic pipeline for branch recovery.

identify a branch misprediction, the speculative processor state and the program counter should be restored to the point where we started processing instructions from the wrong path.

The processor recovery after a branch misprediction is typically split into two separate tasks: front-end recovery and back-end recovery. The front-end recovery is usually simpler than the back-end recovery. In general, recovering the front-end implies flushing all intermediate buffers where instructions fetched from the wrong path are waiting to be renamed, restoring the history of the branch predictor and updating the program counter to resume fetching instructions from the correct path. By contrast, recovering the back-end implies removing all instructions belonging to the wrong path residing on any buffer like the memory order buffer, issue queue, reorder buffer, etc. Moreover, the renaming tables should be restored as well in order to properly rename instructions from the correct path. Finally, back-end resources like physical registers or issue queue entries allocated by wrong-path instructions should also be reclaimed.

Figure 8.3 shows the recovery process on a hypothetical x86 pipeline in the event of a branch misprediction. As commented before, the front-end is recovered earlier than the back-end so that it can start fetching instructions from the correct path early. Then, the back-end recovery is overlapped with the fetch of the first instructions from the correct path. These instructions can flow through the front-end pipeline until the allocation stage. Note that the allocation stage cannot properly handle these new instructions until the renaming tables are fully recovered, and the resources allocated by wrong-path instructions are reclaimed. Therefore, if the front-end pipeline is shorter than the back-end recovery, these instructions are buffered before the renaming stage until the back-end recovery ends.

8.3.1.1 Handling Branch Mispredictions on an ROB-Based Architecture with RRF. ROB-based architectures that implement a retire register file (RRF) could implement any of the recovery techniques that we will describe for merged register files. However, it is also common for these architectures to implement a mechanism similar to the Intel Pentium Pro.

When a branch misprediction is encountered in an Intel Pentium Pro, the processor does not recover the speculative state until all instructions previous to the mispredicted branch and also this branch have been committed. At this point, it is guaranteed that the architectural state in the RRF represents the application state after the execution of the mispredicted branch. Then, the renaming table at the allocation stage is restored by making all its entries to point the values in the RRF in order to begin the renaming of instructions from the correct path.

8.3.1.2 Handling Branch Mispredictions on a Merged Register File. Processors implementing a merge register file do not usually wait for the mispredicted branch to commit in order to recover the speculative state.

These processors keep a log of the way the renaming table is modified when an instruction is renamed and the resources this instruction allocated. Then, in the event of a branch misprediction, this log is traversed in order to recover the correct state by the time the branch was renamed.

This log usually comprises one entry per renamed instruction where every entry contains the following fields: the logical register the instruction overwrites and either the physical register identifier assigned to this instruction or the physical register identifier assigned to the previous writer of the same logical register. The log includes the physical register identifier assigned to this instruction or the previous writer of the same logical register depending on whether the traversal is done forward or backwards.

Traversing this log may take a very long time if there are many instructions in flight to be walked. Therefore, processors like the MIPS R10000 or Alpha 21264 rely on a checkpoint mechanism in order to reduce the distance between the point we start the traversal and the mispredicted branch. These processors periodically take a snapshot of the content of the renaming table so that the log does not have to be fully traversed, but the traversal begins on an instruction where a checkpoint was taken.

For example, in case of a MIPS R10000 processor, the first checkpoint younger than the branch is copied into the renaming table, and the renaming log is traversed backwards until the mispredicted branch is found. Every entry on the log includes the previous mapping of the logical register that the renamed instruction overwrote. Then, the renaming table is restored based on this information in order to reflect the renaming mappings it had by the time the mispredicted branch was renamed. If there is no any valid checkpoint available younger than the mispredicted branch, the traversal begins on the youngest renamed instruction and updates the existing renaming table.

Besides the renaming table, other information like, for instance, the list of available physical register identifiers should be updated to include those registers allocated by the instructions in the wrong path. Some processors like the Alpha 21264 implement the list of free physical registers as part of the checkpoint. Then, this list is restored starting a log traversal from the checkpoint the same way it is done for the renaming table.

8.3.2   RECOVERY FROM AN EXCEPTION

Exceptions are usually handled at commit time. The reason is twofold: first, we need to be sure that the instruction that triggered the exception is not speculative; for instance, it does not belong to a wrong path. Second, we need to provide the architectural state the way it would be as if all instructions previous to this one would have been executed on the original sequential order. Then, all in-flight instructions are flushed because the exception should be handled before resuming the execution of the application. At this point, the speculative state is recovered using one of the mechanisms explained in the previous section. Finally, the front-end is redirected to start fetching instructions from the exception handler.

. . . .

References

[1]   J. Abella, R. Canal, and A. González. Power- and Complexity-Aware Issue Queue Designs. *IEEE Micro*, 23(5):50–58, September–October 2003. doi:10.1109/MM.2003.1240212

[2]   A. Agarwal, and S.D. Pudar. Column-Associative Caches: A Technique for Reducing the Miss Rate of Direct-Mapped Caches. In *Proceedings of the 20th Annual International Symposium on Computer Architecture*, May 1993.

[3]   AMD. 3DNow!™ Technology Manual. March 2000.

[4]   M. Baron. Cortex-A8: High Speed, Low Power. *Microprocessor Report*, November 2005.

[5]   D. Boggs *et al.* The Microarchitecture of the Intel® Pentium® 4 Processor on 90nm Technology. *Intel Technology Journal*, 8(1), February 2004.

[6]   R. Canal, J.M. Parcerisa, and A. González. A Cost-Effective Clustered Architecture. In *Proceedings of the International Conference on Parallel Architectures and Compilation Techniques*, New Port Beach (USA), Oct. 12–16, 1999, pp. 160–168.

[7]   R. Canal, J.M. Parcerisa, and A. González. Dynamic Cluster Assignment Mechanisms. In *Proceedings of the Sixth International Symposium on High-Performance Computer Architecture*, Toulouse (France), Jan. 10–12, 2000, pp. 133–142.

[8]   J. Cruz, A. González, M. Valero, and N. Topham. Multiple-banked register file architectures. In *Proceedings of the 27th Annual International Symposium on Computer Architecture*, 2000.

[9]   J.H. Edmondson *et al.* Internal Organization of the Alpha 21164, a 300-MHz 64-bit Quad-issue CMOS RISC Microprocessor. *Digital Technical Journal*, 7(1), 1995.

[10]   K.I. Farkas, and N. Jouppi. Complexity/Performance Tradeoffs with Non-Blocking Loads. In *Proceedings of the 21th Annual International Symposium on Computer Architecture*, April 1994.

[11]   K.I. Farkas, P. Chow, N.P. Jouppi, and Z. Vranesic. The Multicluster Architecture: Reducing Cycle Time Through Partitioning. In *Proceedings of the 30th Annual International Symposium on Microarchitecture*, December 1997.

[12]   M. Franklin, and G. Sohi. Non-blocking caches for high performance processors, 1991. Manuscript.

[13] G. Gerosa *et al.* A Sub-1W to 2W Low-Power ISA Processor for Mobile Internet Devices and Ultra-Mobile PCs in 45nm Hi-κ Metal Gate CMOS. In *Proceedings of the International Solid-State Circuits Conference*, February 2008.

[14] L. Gwennap. Intel's P6 Uses Decoupled Superscalar Design. *Microprocessor Report*, 9(2), February 1995.

[15] T.R. Halfhill. Intel's Tiny Atom: New Low-Power Microarchitecture Rejuvenates the Embedded x86. *Microprocessor Report*, April 2008.

[16] J.L. Hennessy, and D.A. Patterson. *Computer Architecture: A Quantitative Approach*. 4th Edition. ISBN: 978-0-12-370490-0. Nov. 2006.

[17] IBM. *PowerPC Microprocessor Family: Vector/SIMD Multimedia Extension Technology Programming Environments Manual*. Aug. 2005.

[18] Intel Corp. *Intel® Advanced Vector Extensions Programming Reference*. http://www.intel.com/software/avx.

[19] Intel Corp. *Intel® 64 and IA-32 Architectures. Software Developer's Manual*. Vol. 2A, March 2009.

[20] Intel Corp. *Intel® 64 and IA-32 Architecture. Software Developer's Manual*. Vol. 2B, March 2009.

[21] Intel Corp. *Intel® 64 and IA-32 Architectures Optimization Reference Manual*. March 2009.

[22] T. Juan, J.J. Navarro, and O. Temam. Data Caches for Superscalar Processors. In *Proceedings of the International Conference on Supercomputing*, July 1997.

[23] D. Kanter. Inside Nehalem: Intel's Future Processor and System. Real World Technologies, February 2008. http://realworldtech.com/.

[24] C.N. Keltcher, K.J. McGrath, A. Ahmed, and P. Conway. The AMD Opteron Processor for Multiprocessor Servers. *IEEE Micro*, March/April 2003.

[25] R.E. Kessler, E.J. McLellan, and D.A. Webb. The Alpha 21264 Microprocessor Architecture. In *Proceedings of the International Conference on Computer Design*, October 1998.

[26] R.E. Kessler. The Alpha 21264 Microprocessor. *IEEE Micro*, March/April 1999. doi:10.1109/40.755465

[27] D. Koufaty, and D. T. Marr. Hyperthreading Technology in the Netburst Microarchitecture. *IEEE Micro*, 23(2):56–65, March 2003.

[28] D. Kroft. Lockup-Free Instruction Fetch/Prefetch Cache Organization. In *Proceedings of 8th International Symposium on Computer Architecture*, May 1981.

[29] A. Kumar. The HP PA-8000 RISC CPU. *IEEE Micro*, 17(2):27–32, March 1997. doi:10.1109/40.592310

[30] H.Q. Le, W.J. Starke, J.S. Fields, F.P. O'Connell, D.Q. Nguyen, B.J. Ronchetti, W.M. Sauer,

E.M. Schwarz, and M.T. Vaden. IBM POWER6 Microarchitecture. *IBM Journal of Research and Development*, 51(6), Nov 2007.

[31] S. McFarling. *Combining Branch Predictors*, Technical Note TN-36, DEC Western Research Laboratory, 1993.

[32] M. Prices. *MIPS IV Instruction Set*. Revision 3.2. MIPS Technologies, 1995.

[33] E.J. McLellan, and D.A. Webb. The Alpha 21264 Microprocessor Architecture. In *Proceedings of the International Conference on Computer Design*, p. 90, October 05-05, 1998.

[34] S. Palacharla. Complexity-Effective Superscalar Processors, PhD Thesis, University of Wisconsin-Madison. 1998.

[35] J.A. Rivers, G.S. Tyson, E.S. Davidson, and T.M. Austin. On High-Bandwidth Data Cache Design for Multi-Issue Processors. In *Proceedings of the 30th International Symposium on Microarchitecture*, December 1997.

[36] E. Rotenberg, S. Bennet and J.E. Smith. Trace Cache: A Low Latency Approach to High Bandwidth Instruction Fetching. In *Proceedings of the International Symposium on Microarchitecture*, Paris (France), pp. 24–35, 1996.

[37] A. Seznec. A case for two-way skewed-associative caches. In *Proceedings of the 20th International Symposium on Computer Architecture*, May 1993.

[38] B. Sinharoy, R.N. Kala, J.M. Tendler, R.J. Eickenmeyer, and J.B. Joyner. POWER5 System Microarchitecture. *IBM Journal of Research and Development*, 49(4/5), July/September 2005.

[39] M. Slater. Intel Boosts Pentium Pro to 200 MHz. *Microprocessor Report*, 9(15), November 13, 1995.

[40] J.E. Smith. A Study of Branch Prediction Strategies. In *Proceedings of the International Symposium on Computer Architecture*, pp. 135–148, 1981.

[41] J.E. Smith, and A.R. Pleszkun. Implementing Precise Interrupts in Pipelined Processors. *IEEE Transactions on Computers*, 37(5):562–573, May 1988.

[42] J.M. Tendler, J.S. Dodson, J.S. Fields, Jr., H. Le, and B. Sinharoy. POWER4 System Microarchitecture. IBM *Journal of Research and Development*, 46(1), January 2002.

[43] R.M. Tomasulo. An Efficient Algorithm for Exploiting Multiple Arithmetic Units. *IBM Journal of Research and Development*, pp. 25–33, Jan. 1967.

[44] N. Topham, and A. González. Randomized Cache Placement for Eliminating Conflicts. *IEEE Transactions on Computers*, 48(2), February 1999. doi:10.1109/12.752660

[45] S.W. White. POWER2: Architecture and Performance. In *Proceedings of Spring COMPCON'94 Digest of Papers*, February/March 1994.

[46] White paper. *OpenSPARC™ T1 Microarchitecture Specification*. Sun Microsystems, 2006.

[47] White Paper. *Transmeta™ Efficeon™ TM8600 Processor*. http://datasheets.chipdb.org/Transmeta/pdfs/brochures/efficeon_tm8600_processor.pdf

[48] K.C. Yeager. The MIPS R10000 Superscalar Microprocessor. *IEEE Micro*, April 1996. doi:10.1109/40.491460

[49] T.Y. Yeh, and Y.N. Patt. Alternative Implementations of Two-Level Adaptive Brach Prediction. In *Proceedings of the International Symposium on Computer Architecture*, pp. 124–134, 1992.

[50] V. Zyuban. Inherently Lower-Power High-Performance Superscalar Architectures, PhD Thesis, Univ. of Notre Dame, January 2000.

[51] V.V. Zyuban, and P.M. Kogge. Inherently Lower-Power High-Performance Superscalar Architectures. *IEEE Transactions on Computers*, 50(3), March 2001. doi:10.1109/12.910816

Author Biographies

Antonio González received his Ph.D. degree from the Universitat Politècnica de Catalunya (UPC), Barcelona, Spain, in 1989. He is the founding director of the Intel Barcelona Research Center, started in 2002, whose research focuses on computer architecture. He has been a faculty member of the Computer Architecture Department of UPC since 1986 and became a full professor in 2002.

Antonio has filed over 40 patents, has published over 300 research papers, and has given over 80 invited talks in the areas of computer architecture and compilers. He has served as an Associate Editor of several IEEE and ACM journals, has been a member of the program committee of numerous symposia, the program chair for some of them, including ISCA, MICRO, HPCA, ICS, and ISPASS, and the general chair for MICRO.

Antonio's awards include the award to the best student in computer engineering in Spain graduating in 1986, the 2001 Rosina Ribalta Award as the advisor of the best PhD project in Information Technology and Communications, the 2008 Duran Farrell Award for research in technology, and the 2009 Aritmel National Award of Informatics to the Computer Engineer of the Year.

Fernando Latorre received his M.S. degree from the University of Zaragoza, Spain, in 2001 and his Ph.D. degree from the Universitat Politècnica de Catalunya (UPC), Barcelona, Spain, in 2009. His thesis focused on efficiently exploiting instruction level parallelism and thread level parallelism using adaptive multithreaded/multicore architectures. Fernando joined the Intel Barcelona Research Center in 2003 where he is a Senior Research Scientist, and he is also a member of the Architectures and Compilers research group of the UPC. His research interests range from power-efficient architectures to co-designed virtual machines and parallel processors.

Fernando holds 2 patents, has filed several more, and has published more than 10 research papers in the area of computer architecture. He has served as a reviewer for numerous ACM and IEEE conferences and symposia, and was also a program committee member for WEED 2010 and ISCA 2011. In 2008, he received the Duran Farrell Award for research in technology.

Grigorios Magklis received his B.Sc. degree in Computer Science from the University of Crete, Greece, in 1998. He received his M.Sc. and Ph.D. degrees from the Computer Science Department of the University of Rochester, NY, in 2000 and 2005, respectively. His thesis focused on increasing the energy efficiency of adaptive architectures. In 2003, Grigorios joined the Intel Barcelona Research Center as a Senior Research Scientist. He is also a member of the Architectures and Compilers research group of the Universitat Politècnica de Catalunya, Barcelona, Spain, where he remains until today. His research interests include power-efficient architectures, parallel architectures, dynamic optimization, and operating systems, among others.

Grigorios holds 6 patents and has published more than 20 research papers in the area of computer architecture and distributed systems. He has served as a reviewer for numerous ACM and IEEE conferences and symposia, and was also the architecture track program chair for IPDPS 2009. In 2008, he received the Duran Farrell Award for research in technology.

Printed in the United States
by Baker & Taylor Publisher Services